Nanomaterials for Energy Applications

Topical Conference at the 2009 AIChE Annual Meeting

Nashville, Tennessee, USA
8-13 November 2009

ISBN: 978-1-61567-916-4

Printed from e-media with permission by:

Curran Associates, Inc.
57 Morehouse Lane
Red Hook, NY 12571

Some format issues inherent in the e-media version may also appear in this print version.

Printed by Curran Associates, Inc. (2010)

For permission requests, please contact AIChE
at the address below.

AIChE
3 Park Avenue
New York, NY 10016-5991

Phone: (203) 702-7660
Fax: (203) 775-5177

www.aiche.org

Additional copies of this publication are available from:

Curran Associates, Inc.
57 Morehouse Lane
Red Hook, NY 12571 USA
Phone: 845-758-0400
Fax: 845-758-2634
Email: curran@proceedings.com
Web: www.proceedings.com

TABLE OF CONTENTS

Theoretical Analysis of the Hydrogen Storage Capacity Limitations of Single-Walled Carbon Nanotubes

Andre R. Muniz, Department of Chemical Engineering, University of Massachusetts, Amherst, MA

Dimitrios Maroudas, Department of Chemical Engineering, University of Massachusetts, Amherst, MA

When single-walled carbon nanotubes (SWCNTs) are exposed to a hydrogen plasma, atomic hydrogen is chemisorbed onto their graphene walls. One potential application of this process is the use of SWCNTs as media for hydrogen storage. Experimental studies have reported storage capacities obtained by this process as high as 7 wt %; this is equivalent to a nanotube wall coverage by H atoms of almost 100%, which is the theoretical maximum value. However, in many experiments, lower H storage capacities have been reported consistently and the reasons for this limitation are not well understood. For practical applications of this process, a fundamental understanding is required of the structural changes undergone by SWCNTs upon hydrogenation.

In this presentation, we report results of a computational atomic-scale analysis of the effects of atomic hydrogen chemisorption on the structure and deformation of SWCNTs. The analysis is based on classical molecular-dynamics (MD) and Monte Carlo (MC) simulations of structural and compositional relaxation, as well as targeted first-principles density functional theory (DFT) calculations that complement and validate the classical simulation results. In the MD and MC simulations, the interatomic interactions are described according to the Adaptive Interatomic Reactive Empirical Bond Order (AIREBO) potential. The DFT calculations are performed within the generalized gradient approximation and employ plane-wave basis sets, ultrasoft pseudopotentials, and supercell models.

We find that H chemisorption induces structural changes in SWCNTs associated with sp^2-to-sp^3 bonding transitions, which cause deformation and amorphization of the SWCNT wall. A particularly important effect is the "swelling" of the nanotube, consistently with experimental observations. The corresponding computed radial and axial strains depend on the H coverage and on the SWCNT diameter and chirality. Most importantly, we find that there is a critical H coverage (typically ^{3} 30%), beyond which the radial and axial SWCNT strains start increasing linearly with H coverage and sp^3-hybridized C atoms prevail; at sub-critical coverages, the strain levels are negligible and sp^2-hybridized C atoms dominate. When SWCNTs arranged in bundles are exposed to atomic hydrogen, this swelling effect upon hydrogenation may limit the total amount of hydrogen that can be chemisorbed on the SWCNT walls. The swelling of the SWCNTs results in a decrease in the intertube spacing within the bundle, which may hinder the diffusion of atomic hydrogen through the interstitial space of the bundle. This mass-transfer limitation may cause the non-uniform hydrogenation of the SWCNTs in the bundle and, consequently, a decrease in the total amount of hydrogen that can be chemisorbed in the SWCNTs and stored in the bundles. We present an analytical model for this phenomenon. The model is parameterized according to the atomic-scale computations of SWCNT swelling and provides estimates of the maximum tube wall coverage that can be obtained as a function of the bundle density and the properties of the individual nanotubes in the bundle. The model predictions are assessed by comparisons with large-scale MD/MC simulations of hydrogenation of SWCNT bundles.

See more of this Session: Nanomaterials for Energy Storage I
See more of this Group/Topical: Topical 5: Nanomaterials for Energy Applications

LiMn(BH4)3 for On-Board Hydrogen Storage

Pabitra Choudhury, Chemical and Biomedical Engineering, University of South Florida, Tampa, FL

Sesha S. Srinivasan, Clean Energy Research Center, College of Engineering, University of South Florida, Tampa, FL

Venkat R. Bhethanabotla, Chemical and Biomedical Engineering, University of South Florida, Tampa, FL

Yogi Goswami, Chemical and Biomedical Engineering, University of South Florida, Tampa, FL

Elias Stefanakos, Electrical Engineering, University of South Florida, Tampa, FL

In this study, a promising complex borohydride $LiMn(BH_4)_3$ was evaluated for on-board hydrogen storage. This new complex hydride was prepared from the precursor materials $LiBH_4$ and $MnCl_2$ *via* the solid state mechano-chemical process. The B-H stretch occurrence at 2374 cm^{-1} in addition to two other B-H bonding bands of $LiBH_4$ (2228 and 2297 cm^{-1}) from the FTIR investigation confirm the formation of $LiMn(BH_4)_3$ at room temperature. Gas chromatography studies of both undoped and doped LiMn $(BH_4)_3$ demonstrate that the evolved gas is mainly hydrogen and does not contain members of the borane family. Though this complex $LiMn(BH_4)_3$ exhibits high theoretical hydrogen storage capacity (~8 wt.%) at lower temperature (<150 ^{o}C), the reversible hydrogenation and dehydrogenation cycling is not promising. A strategy has been adopted to convert this new material to become reversible, and these results will also be presented.

See more of this Session: Nanomaterials for Energy Storage I
See more of this Group/Topical: Topical 5: Nanomaterials for Energy Applications

Theoretical Thermodynamic and Kinetic Studies of Hydrogen Desorption in Ti-Doped Sodium Alanates

Gopi Krishna Phani Dathar, Louisiana Tech University, Ruston, LA

Daniela S. Mainardi, Chemical Engineering, Nanosystems Engineering & Institute for Micromanufacturing, Louisiana Tech University, Ruston, LA

Pristine and titanium-doped sodium alanates are one of the pioneering and widely researched materials for hydrogen storage. Limitations related to the high temperatures for release and the slow release rate of hydrogen that are inherent in pristine sodium alanates can be overcome with the use of titanium. This study aims at theoretically understanding the thermodynamic and kinetic profiles of Ti-doped sodium alanates.

The formation of TiAl3 species is evident and stoichiometrically favorable from the atomic compositions of titanium and aluminum in Ti-doped NaAlH4 models. However, the pathways for TiAl3 formation or the intermediate phases that result and take part in dehydrogenation reactions are not completely understood. Six thermodynamic reaction pathways in Ti-doped sodium alanates that lead to the formation of TiAl3 and hydrogen release are studied in this work to propose the most favorable one. Of all the possible reaction pathways, the three-step reaction that proceeds with aluminum hydrides as intermediates in the first, TiAl in the second, and the final product TiAl3 in the third step of the reaction is the most favorable one. This reaction pathway leads to the formation of TiAl3 and aluminum in its +3 oxidation state present in aluminum hydride species responsible for the formation of Ti-Al alloys.

In this work, the free energy barriers associated with the reactions involved in the evolution of hydrogen via the first step of decomposition of pristine sodium alanates (NaAlH4→1/3Na3AlH6+2/3Al+H2) are also studied. Results from our calculations suggest a four step reaction that includes the transition from AlH4- to AlH63- anions, Al clustering and H2 evolution. The free energy barrier associated with one molecule of H2 release lies in the range of 80-124 kJ/mol H2 and the enthalpy of activation lies between 82-120 kJ/mol H2. The rate determining step for this mechanism is found to be the hydrogen evolution from associated AlH3 species. The role of Ti dopants in improved kinetics of Ti-doped sodium alanates is elucidated from our DFT/DFT-MD calculations. Ti dopants stay on the surface; serve as catalytic species in splitting hydrogen from AlH4/AlH3 groups as well as initiators for Al nucleation sites in Ti-doped NaAlH4.

See more of this Session: Nanomaterials for Energy Storage I
See more of this Group/Topical: Topical 5: Nanomaterials for Energy Applications

Chemically Driven, Carbon Nanotube-Guided Thermopower Waves

Won Joon Choi, Chemical Engineering, Massachusetts Institute of Technology, Cambridge, MA

Joel Abrahamson, Chemical Engineering, Massachusetts Institute of Technology, Cambridge, MA

Jae Hee Han, Chemical Engineering, Massachusetts Institute of Technology, Cambridge, MA

Changsik Song, Chemical Engineering, Massachusetts Institute of Technology, Cambridge, MA

Michael Strano, Chemical Engineering, Massachusetts Institute of Technology, Cambridge, MA

Coupling an exothermic chemical reaction with a nanotube or nanowire possessing high axial thermal conductivity creates a self propagating reactive wave along its length. Such waves are realized using a 7 nm cyclotrimethylene-trinitramine (TNA) annular shell around a multi-walled carbon nanotube and are amplified by more than 100 times the bulk TNA value, propagating more than 3 m/s, with an effective thermal conductivity of 1.00 ± 0.35 kW/m/K at 2860 K. This wave produces a concomitant electrical pulse of disproportionately high specific power, as large as 7 kW/kg, that we identify as a thermopower wave. The specific power of such thermopower waves demonstrates an unusual inverse scaling with system size, highlighting their utility as sub-micron and nano-sized pulse power sources. The reaction also produces an anisotropic pressure wave of exceedingly high total impulse per mass of 300 N-s/kg and specific impulse per total mass (5.5 s/μg) having applications to micro-propulsion and actuators.

See more of this Session: Nanomaterials for Energy Storage I
See more of this Group/Topical: Topical 5: Nanomaterials for Energy Applications

Hydrogen Storage in Li-Doped 3D Covalent Organic Borosilicate Framework

Dapeng Cao, College of Chemical Engineering, Beijing University of Chemical Technology, Beijing, China

Jianhui Lan, College of Chemical Engineering, Beijing University of Chemical Technology, Beijing, China

Wenchuan Wang, College of Chemical Engineering, Beijing University of Chemical Technology, Beijing, China

We used the multiscale theoretical method [J. Phys. Chem. C, 2008, 112, 5598], which combines the first-principles calculation and grand canonical Monte Carlo simulation, to investigate the adsorption capacities of hydrogen on non-doped and Li-doped 3D covalent organic borosilicate frameworks (COBFs). Results show that the gravimetric and volumetric storage capacities of hydrogen in Li-doped COBF are 2.83 wt % and 15.80 g/L at 298 K and p=100 bar, showing the enhancement by 76% and 85% than those in non-doped COBF, respectively. It is suggested that to our best knowledge, Li-doped COBF is one of the most promising candidates for hydrogen storage so far.

See more of this Session: Nanomaterials for Energy Storage I
See more of this Group/Topical: Topical 5: Nanomaterials for Energy Applications

Alkyl Functionalized Silicon Surfaces

Sri Sai S. Vegunta, Johnpeter N. Ngunjiri, Wanli Xu and John C. Flake
Louisiana State University, Baton Rouge, LA 70803.
Extended Abstract for Abstract #158540

Introduction

Tailoring the surface chemistry of planar or nanoscale silicon surfaces may enable several new applications such as ultra sensitive biosensor FETs, anodes for lithium ion batteries or new types of solar cells. Silicon naturally forms native oxides (<5nm); however, this layer is typically removed with HF and replaced with furnace oxide or CVD-grown dielectric layer. A relatively stable organic surface with tailored terminal chemistry and controllable electronic properties may be created by functionalizing the surface using chemical or electrochemical methods. Chemical methods typically involve the reaction of silanes with oxides or hydroxides at the silicon surface and electrochemical methods typically result in direct Si-C interfaces (no oxide intermediates). Direct Si-C bonds are relatively inert compared to Si-H bonds and offer less susceptibility to nucleophilic substitution reactions. Electrochemical routes are also interesting to direct silicon surface modification because of relative ease, controllability and effectiveness of the technique.[1,2]

Electrochemical pathways typically include cathodic and anodic grafting of unsaturated organic moieties and anodic grafting of alkyl halides and alkyl Grignards. The cathodic electrografting proceeds via reduction of surface (100) Si-H bonds to obtain silyl anion intermediates, which in turn deprotonates weakly acidic alkyl groups in the organic electrolyte (alkyne/alkene) solution. The silyl anions further undergo nucleophilic substitution reactions with the alkyl carbanions to obtain covalently grafted alkyne monolayer upon hydride terminated (100) silicon surface. On the other hand, anodic electrografting proceeds via oxidation of the Grignard reagents to produce alkyl groups, which in turn attack the surface Si-H bonds producing silyl radicals. Another alkyl group covalently bonds to the silyl species and forms a covalent Si-C alkyl monolayer upon the hydride terminated (100) silicon surface. In our works, phenyl acetylene and hexynoic acid are cathodically electrografted and methyl monolayers are anodically electrografted upon (100) silicon surfaces. The grafted surfaces are characterized by various techniques such as cyclic voltammetry (CV), atomic force microscopy (AFM), Fourier transform infrared spectroscopy (FTIR), X-ray photoelectron spectroscopy (XPS) and electrochemical impedance spectroscopy (EIS) and the results are analyzed. Various applications of the grafted monolayers in major fields such as biosensors (protein sensing), resists for microfabrication, photoelectrochemical cells, and lithium ion storage batteries are discussed.

Experiment

In a typical electrografting process (cathodic and anodic), the substrate silicon surface is used as the working electrode against a platinum counter electrode along with a suitable reference electrode (Ag pseudoreference adjusted to Ag/AgCl). In the cathodic electrografting (CEG) process, phenylacetylene or hexynoic acid groups are attached to the silicon surface by applying a cathodic bias of -2.5 V (vs. Ag/AgCl) in the corresponding phenylacetylene/hexynoic acid solution electrolytes. A supporting electrolyte of 0.1M tetraethyl ammonium perchlorate is used under quiescent conditions. In the anodic electrografting (AEG) process, short chain alkyl (methyl and ethyl) groups are anodically electrografted upon hydride terminated (100) silicon surfaces by applying a suitable anodic bias of +0.1 V (vs. Ag/AgCl) in the presence of corresponding alkyl Magnesium chloride solution electrolyte. The anodic grafting with short chain Grignard reagents doesn't require a supporting electrolyte and is performed at quiescent conditions.

Results & Discussion

The surface chemistry of cathodically grafted phenyl acetylene and hexynoic acid surfaces is previously analyzed via cyclic voltammetry (CV) and atomic force microscopy (AFM).[1] Cyclic voltammetry results indicate a rectifying behavior in which a hysteresis observed in the first cycle is significantly reduced over the successive cycles. AFM measurements of thickness and rms roughness reveal smooth, uniform and dense monolayer films of phenyl acetylene and hexynoic acid upon (100) silicon surfaces. The surface chemistry of anodically grafted surfaces is analyzed via cyclic voltammetry (CV), Fourier transform infrared (FTIR) spectroscopy and atomic force microscopy (AFM).[2] Cyclic voltammetry results provide insights into the oxidation potentials of methyl and ethyl Grignards and show a similar rectifying behavior to the CV results of CEG process. The rectifying behavior indicates the saturation of the silicon surface sites available for the electrografting process because of a maturing monolayer upon the surface. FTIR results suggest the presence of electrografted methyl and ethyl monolayers upon the silicon surface in a crystalline alkyl environment. AFM results indicate electrografted monolayers provide uniform, homogenous monolayers of methyl and ethyl groups upon the silicon surface based on the rms roughness calculations. The AFM results also suggest the absence of adventitiously adsorbed carbon content, which is obtained in case of the thermally grafted surfaces.

The passivation performance of electrografted short chain alkyl termination upon silicon surface is estimated via X-ray photoelectron spectroscopy (XPS) and is compared to the performance of thermally grafted alkyl terminated silicon surfaces. The XPS results as shown in Figure 1, show the limited rise of the SiOx characteristic peak (101-105 eV) in the silicon 2p region in case of the electrografted methyl samples for 45 days in ambient air. This indicates effective passivation of electrografted alkyl silicon surfaces towards the native oxide formation for 45 days in ambient air, which is better than the performance of thermally grafted samples (~30 days) under similar conditions.

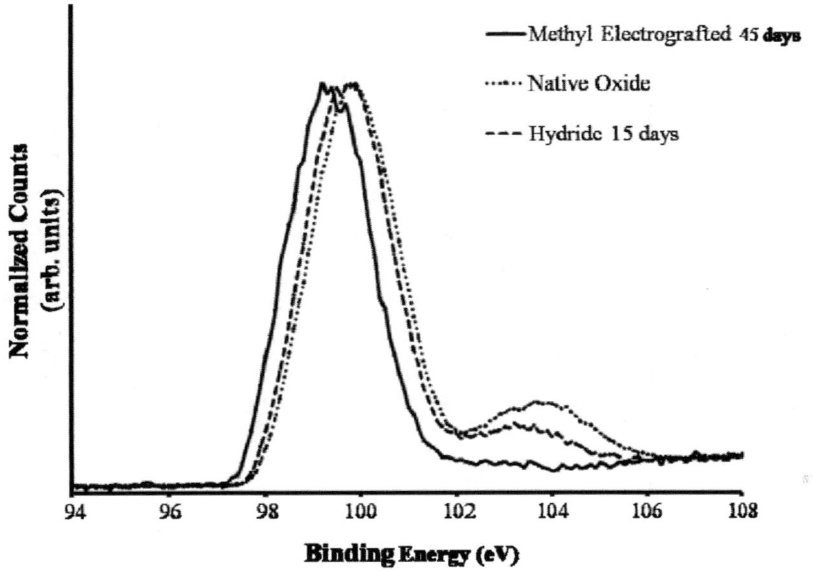

Figure 1: XPS spectra of silicon 2p regions of 45 days old electrografted methyl monolayer compared with 15 days old hydride and native oxide terminated (100) silicon surfaces.

The cathodic grafting results indicate the possibility of functionalizing the silicon surfaces with versatile organic functional groups applicable in major fields of research as enzyme immobilization or biosensors. Likewise, the anodic grafting results suggest short chain alkyl termination upon silicon surfaces provide excellent stability and passivation performance towards native oxide formation for 45 days. This result is the key to many further applications such as monolayer resists, lithium ion batteries or photoelectrochemical cells.

Applications

Cathodic grafting of hexynoic acid upon (100) silicon nanowire surfaces is previously reported for Bovine Serum Albumin (BSA) protein immobilization via EDC chemistry.[3] In a further application, a generalized protein detection technique is introduced by means of the EIS characterization, The grafted hexynoic acid termination upon the silicon surface and the protein (BSA) immobilized hexynoic acid termination upon the silicon surface are subjected to electrochemical impedance (EIS) characterization over a frequency range of 0.05-10 Hz. The protein-immobilized surface is supposed to provide higher impedance due to the thick protein layer upon the silicon surface. As shown in Figure 2, the obtained EIS results are exactly as expected indicating higher impedance for the protein immobilized silicon surface.

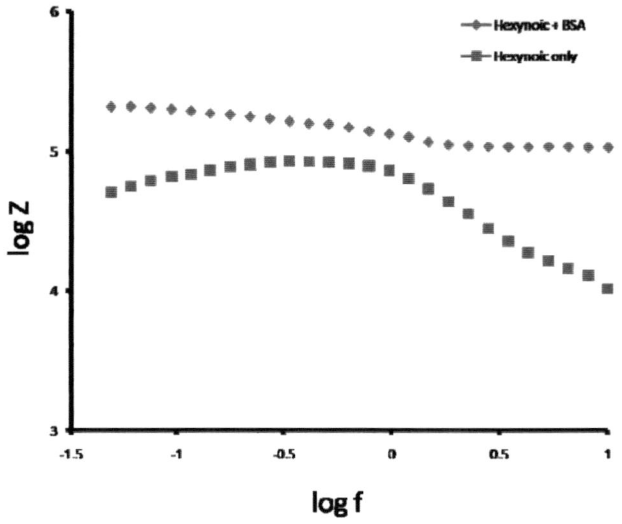

Figure 2: Bode Z impedance plots of Hexynoic acid terminated and BSA immobilized hexynoic acid terminated (100) silicon surfaces.

The difference in the impedance values keeps increasing with increasing frequency indicating the lower double layer capacitance value in the case of BSA immobilized surfaces. This is due to the farther location of the double layer from the semiconductor interface separated by the thick protein layer. The ability of the monolayer surfaces to detect the immobilized proteins via EIS can be further focused towards detection of antibody/antigen pairs.

Patterned templates of cathodically grafted phenyl acetylene and anodically grafted methyl monolayer resists are created using a PDMS stamp in order to show passivation towards electroplating of copper and electrochemical etching of silicon in HF acid solution. Phenyl acetylene monolayer is further characterized by the impedance analysis. The results suggest a higher resistance of the monolayer due to thicker packing density and a lower double-layer capacitance value due to the increased distance from the semiconductor surface by monolayer thickness. Both phenylacetylene and methyl monolayers show excellent resist properties, passivating against copper electrodeposition and electroless etching. The application of the organic films as monolayer resists shows the ability of electrografted organic monolayers to passivate the substrate surface towards the native oxide formation and other applied external forces.[4] Similar major applications of functionalization of silicon surfaces are under research, especially in the fields of lithium ion batteries and photoelectrochemical cells.

References:

1. Ngunjiri J., Vegunta S. S. S., Flake J. C., "Cathodic Electrografting of Alkyl Nanopatterns on Silicon (100)", Journal of Electrochemical Society, 2009, 156 (7), H516-H521.
2. Vegunta S. S. S., Ngunjiri J., Flake J. C., "Electrochemical and Thermal Grafting of Alkyl Grignard Reagents onto (100) Silicon Surfaces", Langmuir, 2009, article in press.
3. Scheibal, Z. R.; Xu, W.; Audiffred, J. F.; Henry, J. E.; Flake, J. C., "Protein immobilization onto silicon nanowires via electrografting of hexynoic acid", Electrochemical and Solid State Letters 2008, 11, (8), K81-K84.
4. Ngunjiri J., Vegunta S. S. S., Flake J. C., "Electrografting of Organic Monolayer Resists onto (100) Silicon", manuscript yet to be submitted.

The Use of Nanoporous Silicon for Advanced Rechargeable Lithium-Ion Batteries

Madhuri Thakur, Department of Chemical and Biomolecular Engineering, Rice Univvresity, Houston, TX

Steven Sinsabaugh, Lockheed Martin Corporation, Akron, OH

Mark Isaacson, Lockheed Martin Corporation, San Jose, CA

Michael Wong, Department of Chemical and Biomolecular Engineering , Department of Chemistry, Rice Univvresity, Houston, TX

Sibani Lisa Biswal, Department of Chemical and Biomolecular Engineering, Rice Univvresity, Houston, TX

There is great interest in developing rechargeable lithium batteries with higher energy capacities and a longer cycle lifetimes. Currently technology has been mainly based on graphite, which has a lithium capacity of 372 mAh/g. Silicon has very high lithium ion capacity of 4200 mAh/g, and has been suggested as a potential high capacity lithium battery anode material. However, it has been encumbered with serious expansion/ contraction during cycling, increasing internal resistance and resulting in cell failure. Severe silicon pulverization can be triggered by a large volume change (>300%) during lithium alloying (to form LixSi) and de-alloying (to reform Si), which results in electrically disconnected smaller particles. These disconnected particles cause a rapid decrease in cycling stability. Porous silicon structures have been suggested as a potential means to accommodate silicon expansion/contraction while maintaining continuity. We will report on our fabrication of a nanoporous silicon structure by electrochemical etching process. The pores vary from 500 nanometer to 1 micrometer in diameter with a pore depth of 10-50 micrometers. A battery cell is created with these nanoporous silicon samples and the battery is tested using charge and discharge rates between 0.1 to 2.5V with a current of 60 microampere. We find that the specific capacity of the battery cell is 1000Mah/g after 10 cycles with a capacity of 2300 mAh/g for first cycle. This leads to the fact that porous silicon can be utilized as a promising electrode for lithium ion battery technology.

See more of this Session: Nanomaterials for Energy Storage II
See more of this Group/Topical: Topical 5: Nanomaterials for Energy Applications

Crystalline MnO2 Nanodomains in Mesoporous Carbon for Electrochemical Capacitors

Keith P. Johnston, Dept. Chemical Engineering, University of Texas at Austin, Austin, TX

Brian Wilson, Dept. Chemical Engineering, University of Texas at Austin, Austin, TX

Domingo Ferrer, Microelectronics Research Center, University of Texas at Austin, Austin, TX

Keith J. Stevenson, Dept. Chemistry, University of Texas at Austin, Austin, TX

Xiqing Wang, Oak Ridge National Lab, TN

Sheng Dai, Oak Ridge National Lab, TN

MnO2/mesoporous carbon hybrid nanocomposites were synthesized to achieve high values of MnO2 capacitance even at high scan rates of 100 mV/s. High resolution transmission electron microscopy (HRTEM) with energy dispersive X-ray spectroscopy (EDX) illustrated the MnO2 was uniformly distributed throughout the mesoporous carbon structure. The highly loaded nanocomposites contained ~1 nm thick MnO2 nanodomains inside of the mesoporous carbon pores. HRTEM and X-ray diffraction (XRD) showed formation of MnO2 nanocrystals with lattice planes corresponding to birnessite. The electrochemical capacitance of these hybrid materials in aqueous 1 M Na2SO4 electrolyte containing as little as 2 wt % MnO2 materials exhibit a high specific capacitance with respect to MnO2 of ~560 F/gMnO2. Even for 30 wt% MnO2, a high MnO2 capacitance of 137 F/gMnO2 on carbon was observed at 100 mV/s. Sodium ion diffusion coefficients on the order of 10-9-10-10 cm2/s were measured with chronoamperometry. Controlled growth of the thickness and crystallinity of conformal, redox-active MnO2 on mesoporous carbon with high surface areas will facilitate high rate, high energy storage.

See more of this Session: Nanomaterials for Energy Storage II
See more of this Group/Topical: Topical 5: Nanomaterials for Energy Applications

Lithium Ion Batteries for Energy Storage Applications: Modeling Diffusion-Induced Stresses in Nanostructured Electrode : Modeling Diffusion-Induced Stresses in Nanostructured Electrode

Rutooj D. Deshpande,Chemical and Materials Engineering, University Of Kentucky, Lexington, KY

Yang -Tse Cheng, Chemical and Materials Engineering, University Of Kentucky, Lexington, KY

Mark W. Verbrugge, Materials and Processes Laboratory, General Motors Research and Development Center, Warren, MI

Developing lithium ion batteries with high power and energy density for wide range of applications is currently under intense research worldwide. Mechanical degradation of the battery electrodes due to lithium ion diffusion during charging and discharging operations is a major hurdle in development of durable lithium ion battery. When lithium is inserted into either the positive or negative electrode, a large volume change on the order of a few to several hundred percent can occur. This volume change due to diffusion induced stresses (DISs) leads to mechanical degradation of the electrode material. Our work is aimed at developing nanostructured electrodes with enhanced durability. Upon decreasing size of the electrode particles, the ratio of surface area to volume increases tremendously. Thus, surface energy and surface stress may play an important role in mitigating DISs in nanostructured electrodes. In this work, we establish relationships between the surface energy, surface stress, and the magnitude of DISs in nanowires. We show that DISs, especially the tensile stresses, can decrease significantly due to the surface effects. Our model also establishes a relationship between stress and the nanowire radius. We show that, with decreasing size, the electrode material will be less prone to mechanical degradation, leading to an increase in the life of lithium ion batteries

See more of this Session: Nanomaterials for Energy Storage II
See more of this Group/Topical: Topical 5: Nanomaterials for Energy Applications

Resorcinol-Formaldehyde Based Electrosprayed Carbon Nanospheres: A Novel Anode Material for Lithium Ion Batteries

Chandra S. Sharma, Suman Saurabh, Sandip Patil, and Ashutosh Sharma

Department of Chemical Engineering and DST Unit on Nanosciences, Indian Institute of Technology, Kanpur-208016, U.P., INDIA

Abstract

Organic resorcinol formaldehyde (RF) sol was electrosprayed to form RF nanodroplets which were collected on a Si wafer. After oven drying at 60°C for 12 hours, RF nano-droplets were pyrolyzed at 900°C in an inert atmosphere to yield the carbon nanospheres. This study reports the optimization of various process parameters including needle diameter, applied electric potential and liquid flow rate in order to get spherical, mono-disperse particles. For the organic RF sol, the optimized parameters, needle diameter 0.241 mm, electric potential 1.5 kV/cm and a flow rate of 0.8 ml/h, enabled the synthesis of nearly monodisperse carbon nano-spheres with diameter of 30.2 ± 7.1 nm. Water contact angle (WCA) studies revealed that the surface properties were significantly influenced by the surface morphologies. The surface covered with the RF derived carbon nano-spheres was super hydrophilic (WCA 10.1°) as compared to a mild hydrophilic nature of the RF derived smooth carbon films (WCA 83.3°). The electrosprayed carbon nanospheres synthesized here may find potential applications as adsorbents and drug carriers, and in carbon-based microelectromechanical systems (C-MEMS) including bio-MEMS.

1. Introduction

Synthesis of spherical carbon particles and other structures of carbon have received a considerable attention in last decade because of their potential use in various fields. Uses of carbon particles are presently being developed in broad ranging applications including anode materials in secondary lithium ion batteries, fuel cells and supercapacitors, adsorbents and drug delivery carriers. Although, there are various approaches reported in the literature to synthesize carbon particles. Some of these include chemical vapor deposition, sol-gel emulsification, and templating techniques. However, despite the fact that there is a significant literature on the synthesis of spherical carbon particles, reports on preparation of carbon nanospheres are very few.

This study establishes electrospraying to be a facile technique for the production of monodisperse carbon nanospheres. In this technique, liquid flowing out from a nozzle is atomized by applying high electric potential. Beyond a threshold value of electric potential, the liquid meniscus elongates to form a jet which finally disrupts into fine charged droplets. Electrospraying offers several advantages over other techniques to form nanoparticles which include synthesis of smaller size particles with a narrow size distribution and high yield.

In this study, we electrosprayed an organic polymer precursor resorcinol-formaldehyde (RF) sol which was first introduced by Pekala (1989) to form carbon microparticles by emulsification. RF based carbon gels have been widely used as adsorbents, anode materials in energy storage devices.There are few reports to form carbon microspheres by preparing a micro-emulsion of RF sol in an organic solvent in the presence of surfactants followed by pyrolysis in

inert atmosphere. However, in this study, we have combined RF sol-gel processing with electrospraying technique to synthesize nearly monodispersed carbon nanospheres.
Resorcinol-formaldehyde based organic sol is successfully electrosprayed for the first time to yield RF nanospheres followed by pyrolysis in an inert atmosphere to obtain carbon nanospheres. Our study shows that the size and morphology of these carbon nanoparticles may be manipulated by varying the process parameters such as needle gauge diameter, applied electric potential and flow rate.
These carbon nanospheres may be utilized in a host of scientific and technical applications including biosensors, adsorbents and anode materials in lithium ion secondary batteries to enhance specific capacity of the electrodes.

2. Experiment

2.1 Materials

Resorcinol (99% purity), formaldehyde (37% w/v; stabilized by 11-14 wt. % methanol), hydrochloric acid (36%) and acetone (99.5%) were purchased from Qualigens Fine Chemicals, India and used as such.

2.2 Preparation of RF Sol

Measured quantities of resorcinol and formaldehyde were added in a beaker and mixed for about 15min by magnetic stirrer to prepare clear solution. For organic RF sol, an acidic catalyst, hydrochloric acid, was added to acetone (used as diluent) separately. The two solutions were then mixed and stirred continuously for 15min. The resorcinol to formaldehyde (R/F) and resorcinol to diluent (R/D) molar ratios were kept constant to be 0.50 and 0.037 respectively. The resorcinol to catalyst molar ratio (R/C) was fixed to be 10 for organic RF sol.

2.3 Experimental Set Up

A schematic diagram of the electrospraying set up used for this work is shown in Fig. 1. It consists of mainly three components: a high voltage DC power supply (Gamma high voltage research Inc., USA), a syringe pump (Longer pump, China) mounted with a needle and a grounded collector plate. The flow rate of the polymer solution can be measured and controlled by syringe pump.

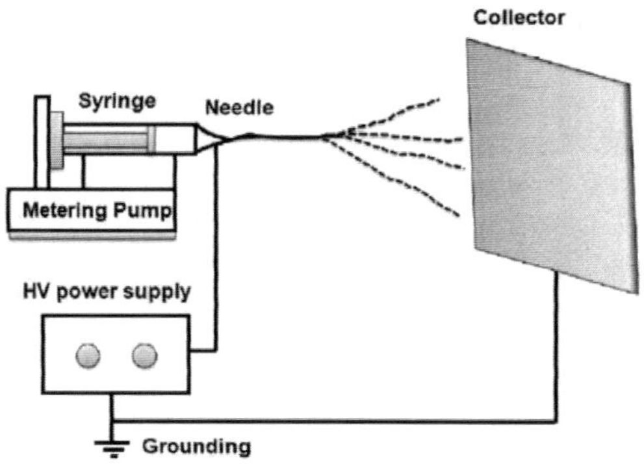

Fig. 1 Schematic Diagram of Electrospraying Set-up

2.4 Preparation of RF sol based Nanospheres

RF based nanospheres were synthesized using the electrospraying process. The organic RF sol just before the onset of gelation was placed in a stainless steel needle (nozzle) containing glass syringe. An external electric field was applied on the sol through needle. Increase in the electric field caused the charges to overcome the surface tension and a straight jet gets ejected from the needle. This jet finally disrupts into the charged droplets while traveling form the tip of the capillary towards the collector. The samples were collected on silicon wafer attached with the grounded copper screen. Samples were than dried by heating in an oven at 333K for 12h.

2.5 Preparation of Carbon Nanospheres

After drying, RF based nanodroplets deposited on a silicon wafer were placed in a quartz boat and pyrolyzed at 1173K in nitrogen (N_2) atmosphere in a tubular, high temperature furnace. Before the pyrolysis process started, nitrogen gas with 0.5 l/min flow rate was purged for about 15 min into the quartz tube to remove unwanted air or oxygen. The rate of heating was optimized to be at 5.0 K/min while the N_2 gas flow during the heating was kept constant at 0.2 l/min. Once the maximum temperature was reached, it was kept constant for 60 min. The furnace was then cooled to room temperature to obtain RF derived carbon nanospheres.

2.6 Characterization of Carbon Nanospheres

The size, morphology and surface roughness of the RF based carbon nanoparticles were studied using scanning electron microscopy (SEM) (SUPRA 40VP, Gemini, Zeiss, Germany). These images were then analyzed using ImageJ software to measure the particle size distribution (PSD). Raman spectra of the pyrolyzed RF based carbon nanospheres was studied by a confocal micro-Raman microscope (CRM 200, WiTec, Germany with λ = 543nm) to characterize the types of bonds between the elements constituting the material.

Wetting behavior of surfaces covered with carbon nanoparticles were characterized by measuring equilibrium water contact angle by contact angle goniometer (Rame-Hart Instruments Co.,NJ, USA). In most of the measurements, we used 5μl (~ 3mm spherical drop diameter) water droplets.

3. Results and Discussion

Micro-Raman spectroscopy was used to confirm the yield of carbon nanospheres upon pyrolysis. A typical Raman spectrum of the pyrolyzed nanoparticles (Fig. 2) shows two broad peaks centered at about 1338 and 1586cm^{-1}. These two peaks correspond to well known D and G bands, associated with the vibrations of sp^2 carbon atoms with dangling bonds. The ratio of intensity of the two first-order D and G band as denoted by I_D/I_G is measured to be 1.72, which indicates a higher fraction of disordered amorphous carbon.

An EDAX spectrum as shown in Fig. 3 also confirms the high yield of carbon (85.3%) after pyrolysis. Some traces of silicon are due to the Si wafer substrate on which these carbon nanospheres were collected during electrospraying.

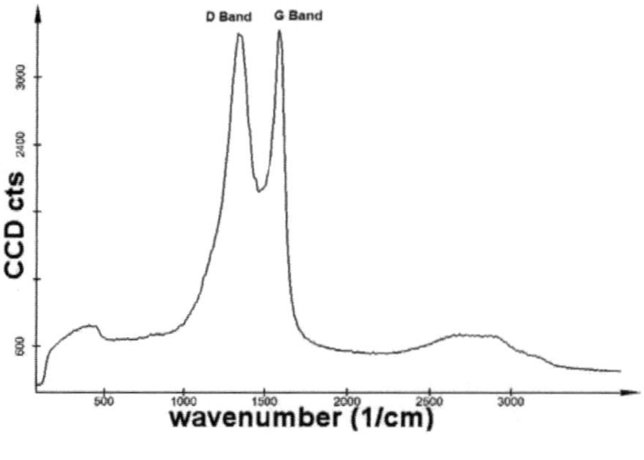

Fig.2 Raman Spectrum of RF sol derived Electrosprayed Carbon Nanospheres

Element	Weight %	Weight % σ	Atomic %
Carbon	85.259	0.391	89.336
Oxygen	11.989	0.400	9.431
Silicon	2.751	0.048	1.233

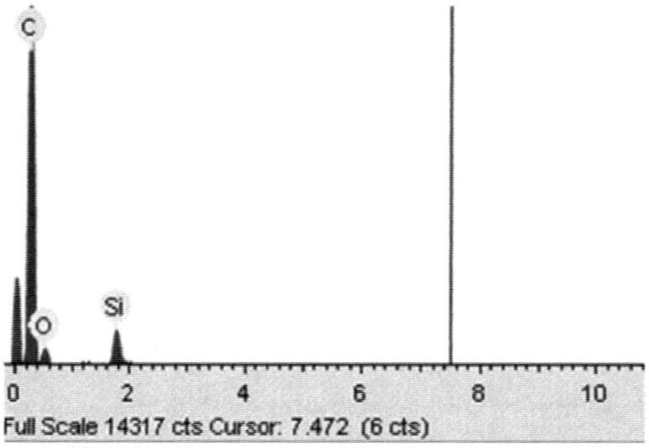

Fig. 3 EDAX Spectrum of Carbon Nanospheres

3.1 Optimization of Process Parameters

3.1.1 Needle Diameter

The effect of needle diameter on particle size and morphology was studied using syringe needles with three different gauges 18, 22 and 26 g (with internal diameter of 0.838, 0.394 and 0.241mm, respectively). Results are summarized here in Fig. 4. Other parameters used were as follows: electric potential 2.0 kV/cm and flow rate 0.8 ml/h. At needle gauge of 18g, we observe a bimodal distribution of RF droplets (Fig. 4a). The average diameter of primary carbon nanoparticles is 156 ± 41.2 nm, while the secondary droplets comprising a smaller fraction have an average diameter of 32.1 ± 4.7 nm.

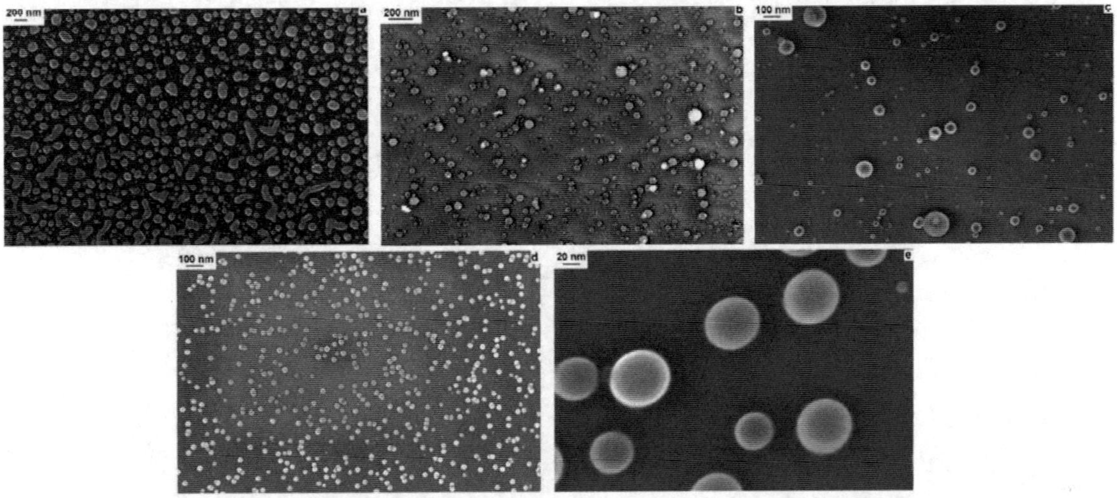

Fig. 4 SEM images showing the effect of needle gauge on the size and morphology of carbon nanospheres: needle gauge (a) 18g; (b) 22g; (c) magnified view of (b); (d) 26g; (e) magnified view of (d). Other parameters are electric potential 2.0kV/cm and flow rate 0.8ml/h.

When employing a narrower needle of 22g, average diameter of carbon nanoparticles was found out to be 74.3 ± 14.2nm (Fig. 4b-c). Further, the jet using 26g needle was even finer than that of 22g needle. It resulted in nearly monodisperse carbon nanospheres with an average diameter of 42.8 ± 6.3 nm as shown in Fig. 4d-e.

As particles using needle gauge 26 were smallest and monodisperse in size and of spherical morphology, syringe needle of 26 g was used for further experiments.

3.1.2 Applied Electric Potential

The effect of applied electric potential on average diameter of carbon nanospheres is summarized in Fig. 5a-h. The electric potential was varied from 1.0 to 2.5 kV/cm with increments of 0.5 kV/cm. Volume feed rate of RF sol was fixed at 0.8 ml/h. At 1.0 kV/cm, larger diameter droplets (average diameter 206.8 ± 26.9nm) were formed as shown in Fig. 5a-b. After an increase in the electric potential to 1.5 kV/cm, there is a significant reduction in the average diameter (36.9 ± 8.1nm) of carbon nanospheres. In these two cases, particles were observed to be nearly monodisperse. On increasing the applied electric potential further to 2.0 kV/cm, nanoparticles with a reduced average diameter, but a much wider size distribution of 23.1 ± 16.7nm were formed as shown in Fig. 5e-f. We observed a prominent bimodal distribution of carbon nanoparticles at very high electric potential (2.5 kV/cm). The average diameter of primary carbon nanoparticles in this case was 43.7 ± 14.9 nm, while the secondary droplets size was 1.4 ± 4.0 nm.

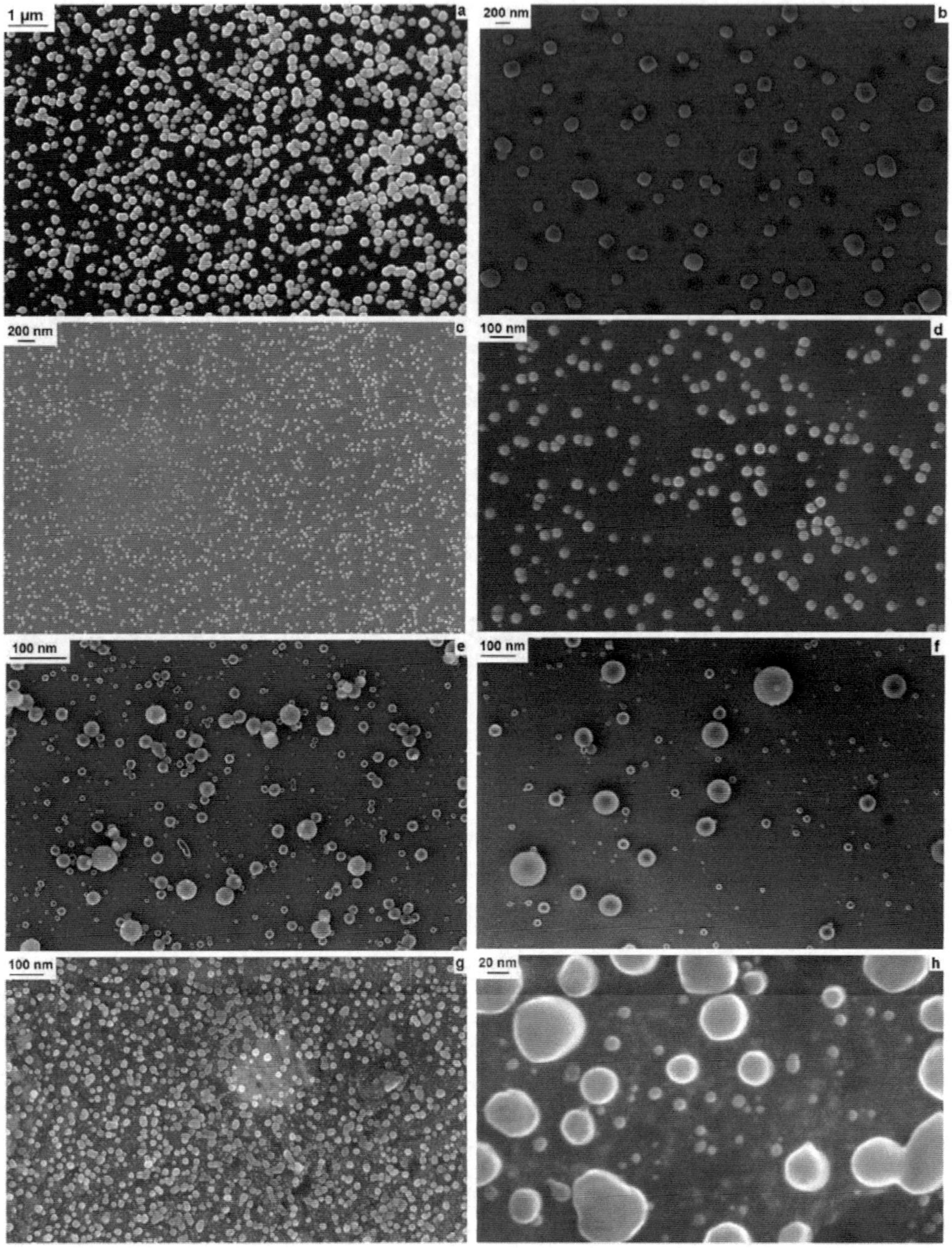

Fig. 5 SEM images showing the effect of applied electric potential on the size and morphology of carbon nanospheres: electric potential (a) 1.0kV/cm; (c) 1.5kV/cm; (e) 2.0kV/cm; (g) 2.5kV/cm. Images (b), (d), (f) and (h) are magnified view of (a), (c), (e) and (g) respectively. Other parameters are needle gauge 26g and flow rate 0.8ml/h.

3.1.3 Volumetric Flow Rate

RF sol flow rate was varied between 0.2 to 2.0 ml/h with increments of 0.6 ml/h to study its effect on nano-particle formation. As shown in Fig. 6a, at 0.2 ml/h, average diameter of particles is 17.1 ± 8.8nm, which increases to 30.2 ± 7.1 nm at 0.8 ml/h (Fig. 6b). In the latter case, carbon nanospheres are nearly monodisperse as well. On increasing the flow rate to 1.4 ml/h, as shown in Fig. 6c-d, average particle diameter of carbon nanospheres was measured to be 88.4 ± 30.1 nm. Further increase in the flow rate to 2.0 ml/h, much larger particles with average diameter 212 ± 37.9 nm are formed which also agglomerate more readily as shown in Fig. 6e.

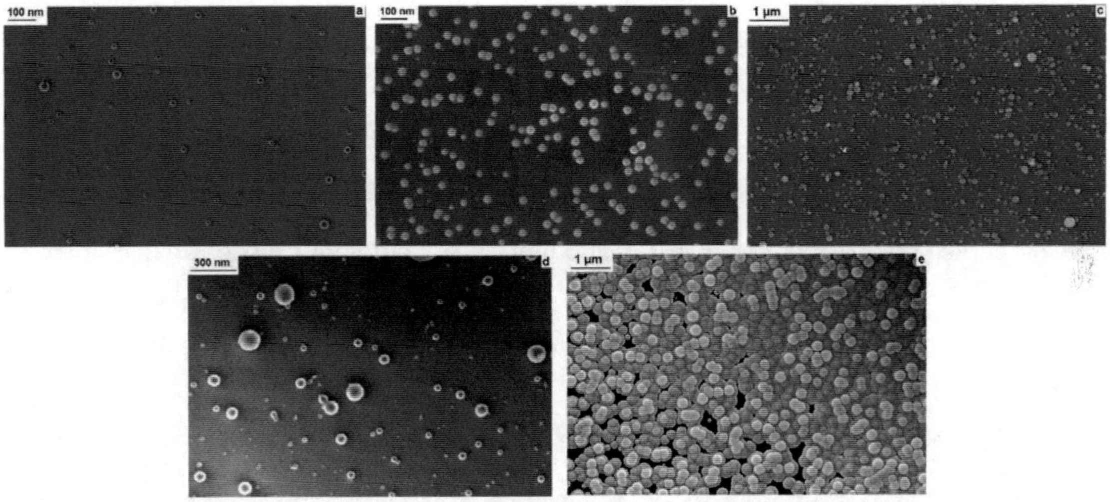

Fig. 6 SEM images showing the effect of volumetric flow rate on the size and morphology of carbon nanospheres: flow rate (a) 0.2ml/h; (b) 0.8ml/h; (c) 1.4ml/h; (d) magnified view of (c); (e) 2.0ml/h. Other parameters are electric potential 2.0kV/cm and flow rate 0.8ml/h.

Based on the studies presented above, needle diameter, applied electric potential and flow rate of RF sol were found to play a key role in controlling the size and morphology of electrosprayed carbon nanospheres. The following set of parameters, needle gauge 26g, applied electric potential 1.5kV/cm at RF sol flow rate 0.8ml/h yielded nearly monodisperse carbon nanospheres with average diameter 30.2 ± 7.1 nm.

3.2 Wettability of surface covered with electrosprayed carbon nanospheres

Wetting behavior of surface covered with electrosprayed carbon nanospheres reported in this work may be useful to find particular applications for these carbon nanospheres. For this, water contact angle was measured on different surfaces which revealed that surface properties were significantly influenced by the surface morphologies. Smooth thin films of organic RF sol were prepared by spin coating it at 3000 rpm and drying at 60°C for 12h. Carbon films were obtained after pyrolysis at the same conditions as mentioned earlier.

Fig. 7 summarizes the contact angle measurement studies on RF sol derived smooth carbon surfaces and on surface covered with carbon nanospheres. The RF sol derived smooth carbon films show a mild hydrophilic behavior (WCA 83.3 ± 1.6°) which is because of the presence of hydroxyl group (-OH group) and other oxygen-containing functional group such as carbonyl as studied by FTIR spectrum (see Reference 10).

Interestingly, the water contact angle on carbon nanosphere surfaces was measured to be 10.1 ± 2.8°, which further reduced to 5 ± 1.8° after 8 min. This observation shows a key role of surface roughness and macro porosity contributed by the nanospheres. By deposition of electrosprayed carbon nanospheres, i.e, by increasing the surface roughness at macro-scale, carbon surface covered with nanospheres approached towards super hydrophilic behavior.

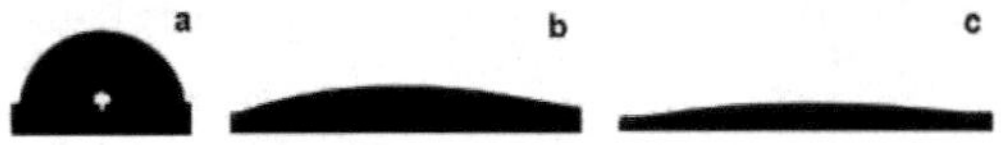

Fig. 7 Water contact angles on: (a) RF sol derived smooth carbon thin film; (b) RF sol derived carbon nanospheres film initially; (c) RF derived carbon nanospheres film after 8 min.

Surface properties and wettability play an important role for many applications of carbon surfaces. Strongly hydrophilic carbon nanosphere surfaces may have applications in biosensors, relative humidity sensing and gas sensing. The carbon nanospheres are also potential candidates in applications involving controlled drug delivery and as an electrode material in rechargeable batteries, fuel cells and super-capacitors. Our group is currently working on the integration of these carbon nanospheres with carbon microelectrochemical systems (C-MEMS) to enhance their performance. Conformal deposition of carbon nanospheres on high aspect ratio carbon posts as shown in schematic diagram below should greatly increase the external surface area available for intercalation of lithium ions thus increasing the specific capacity of the battery electrode.

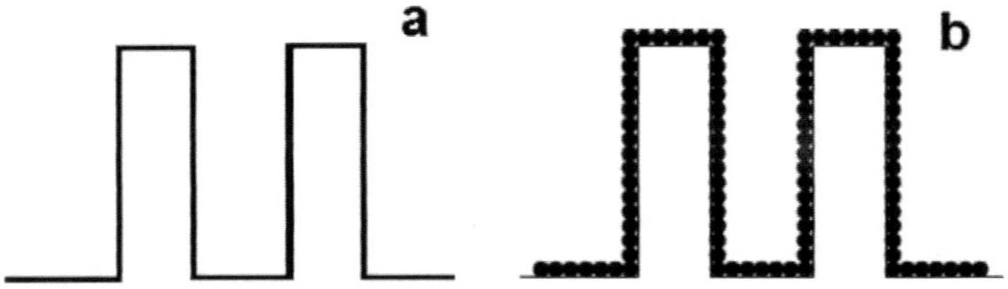

Fig. 8 Schematic of (a) high aspect ratio carbon posts; (b) conformal deposition of carbon nanospheres on high aspect ratio carbon posts

4. Conclusions

This study demonstrates electrospraying to be a facile technique for the production of nearly monodisperse carbon nanospheres. Electrospraying of resorcinol-formaldehyde based organic sol yielded RF nanospheres followed by pyrolysis in an inert atmosphere to obtain carbon nanospheres. The size and morphology of these carbon nanoparticles may be controlled by varying the process parameters such as needle diameter, applied electric potential and volumetric flow rate. Nearly monodisperse carbon nanospheres with average diameter 30.2 ±

7.1nm were obtained by optimizing these parameters. Deviation from the optimized values leads to remarkable changes in the particle size distribution and morphology as discussed. Raman spectroscopy confirmed the amorphous nature of carbon while EDAX analysis also confirmed the higher yield of carbon from resorcinol-formaldehyde based sol. Interesting wettability behavior is observed with carbon nanoparticles surfaces which show extreme structural hydrophilicity (water contact angle ~ 5-10^0) as compared to smooth RF sol derived carbon films (water contact angle ~ 90^0). The carbon nanospheres synthesized here may find suitability in a variety of potential applications including as adsorbents, drug carriers, and electrode material for batteries, super-capacitors and fuel cells.

Acknowledgement

This work is supported by Unilever Research (India), DST Unit on Nanosciences at IIT Kanpur, and by a grant from the DST-IRHPA. Helpful discussions with Marc Madou (UCI) are gratefully acknowledged.

References

1. Alviso C T, Pekala R W, Gross J, Lu X, Caps R and Fricke J 1996 Mat. Res. Soc. Symp. Proc. 431 521.
2. Arya N, Chakraborty S, Dube N and Katti D S 2009 J. Biomed. Mater. Res. Part B: Appl. Biomater. 88B 17.
3. Hasegawa T, Mukai S R, Shirato Y and Tamon H 2004 Carbon 42 2573.
4. Jaworek A, 2007 Powder Technol. 176 18.
5. Jaworek A and Sobczyk A T 2008 J. Electrostatics 66 197.
6. Macdonald S M, Szot K, Niedziolka J, Marken F and Opallo M 2008 J. Solid State Electrochem. 12 287.
7. Pekala R W 1989 J. Mater. Sci. 24 3221.
8. Shanmugam S and Gedanken A 2006 J. Phys. Chem. B 110 2037.
9. Sharma C S, Kulkarni M M, Sharma A and Madou M 2009 Chem. Eng. Sci. 64 1536.
10. Sharma C S, Patil S, Saurabh S, Sharma A and Venkataraghavan R 2009 Bull. Mater. Sci. 32 239.
11. Szot K, Lesniewski A, Niedziolka J, Jonsson M, Rizzi C, Gaillon L, Marken F, Rogalski J and Opallo M 2008 J. Electroanal. Chem. 623 170.
12. Tzeng Y, Huang T S, Chen Y C, Liu C and Liu Y K 2004 New Diamond and Frontier Carbon Tech. 14 193.
13. Wang C, Zaouk R and Madou M 2006 Carbon 44 3073.
14. Wang Y, Serrano S and Santiago-Aviles J J 2003 Synth. Metals. 138 423.
15. Wang Y, Su F, Wood C D, Lee J Y and Zhao X S 2008 Ind. Eng. Chem. Res. 47 2294.

Electrochemical Characterization of P3HT Nanostructures Fabricated Via Template Wetting

Ahmed Minhas, Institute for Manufacturing, Louisiana Tech University, Ruston, LA

Peter Venema, North Carolina State University, Raleigh, NC

Joseph Cannon, Institute for Manufacturing, Louisiana Tech University, Ruston, LA

Steven D. Bearden Jr., Institute for Micromanufacturing, Louisiana Tech University, Ruston, LA

Scott A. Gold,Institute for Micromanufacturing, Louisiana Tech University, Ruston, LA

Polythiophenes have been widely investigated for their potential applications in electrochemical supercapacitors due to their ability to be both n and p doped. In this work we examine poly-3-hexylthiophene, P3HT. Soxhlet extraction was used to remove impurities and to obtain narrow molecular weight fractions from the as received polymer. Template wetting nanofabrication was used with 100 and 200 nm diameter pore, porous alumina membranes to fabricate high surface area electrodes. These were then characterized using chronopotentiometry, cyclic voltammetry, and electrochemical impedance spectroscopy. Specific capacitances of greater than 400 F/g have been achieved with this approach.

See more of this Session: Nanomaterials for Energy Storage II
See more of this Group/Topical: Topical 5: Nanomaterials for Energy Applications

A Cost-Effective Supercapacitor Material of Ultrahigh Specific Capacitances: Spinel Nickel Cobaltite Aerogels From An Epoxide-Driven Sol-Gel Process

Te-Yu Wei, Chemical Engineering, National Tsing-Hua University, Hsin-Chu, Taiwan

Chun-Hung Chen, Chemical Engineering, National Tsing-Hua University, Hsin-Chu, Taiwan

Hsing-Chi Chien, Chemical Engineering, National Tsing-Hua University, Hsin-Chu, Taiwan

Shih-Yuan Lu, Department of Chemical Engineering, National Tsing-Hua University, Hsin-Chu, Taiwan

Chi-Chang Hu, Chemical Engineering, National Tsing-Hua University, Hsin-Chu, Taiwan

Low cost mesoporous materials of high specific surface area, porosity, electronic conductivity, and electrochemical activity, and multiple oxidation states/structures are desired for next generation supercapacitors. These requirements for material characteristics can be met by the nickel cobaltite (NiCo2O4) aerogels prepared from an epoxide-driven sol-gel process. In this work, nickel cobaltite aerogels of ultrahigh specific capacitance (1,400 F/g at a relatively high mass loading of 0.4 mg/cm2, a sweep rate of 25 mV/s, a potential window of 0.04 - 0.52 V in a 1M NaOH solution), excellent reversibility, and outstanding cycle stability were synthesized with a chloride based epoxide addition procedure for the first time. The post-gel-drying calcination temperature was found to play a critical role in producing the preferred products. Nickel cobaltite aerogels of outstanding supercapacitive properties were obtained with a starting Ni/Co ratio of 0.5 and post-gel-drying calcination temperature of 200 oC, achieving an optimal combination of composition, crystallinity, specific surface area, pore volume, and pore size. The development of the present work makes possible the low cost production of next generation, ultrahigh performance supercapacitors of the asymmetric type. The content of this paper is to appear in Adv. Mater.

See more of this Session: Nanomaterials for Energy Storage II
See more of this Group/Topical: Topical 5: Nanomaterials for Energy Applications

Improved Zinc Gallium Oxy-Nitrides as Visible Light Photocatalysts for Hydrogen Production

Venkata Bharat Ram Boppana, Center for Catalytic Science and Technology, Department of Chemical Engineering, University of Delaware, Newark, DE

Raul F. Lobo, Center for Catalytic Science and Technology, Department of Chemical Engineering, Univesity of Delaware, Newark, DE

Photocatalysis is the acceleration of a slow reaction by a photon-assisted process with titanium dioxide (TiO2) being the material of choice for most environmental applications including hydrogen production by water splitting [1,3]. Inspite of a number of advantages like optimum band structure, high surface areas etc, it's efficacy is reduced from the inability to use light with energies less than 3 eV (visible light). The challenge thus lies in the design and development of effective photocatalysts with band gaps less than 3 eV and with efficiencies comparable to TiO2. In the last decade there has been a lot of progress in this direction, introduction of N orbitals increases the valence band level thereby reducing the band gap [2]. Wurzite zinc gallium oxy nitrides (ZGONs) are among the many N substituted compounds that have been investigated as a solution to TiO2 limitations. It has been reported that these ZGONs can produce 1000 μmol/hr of H2 from H2O upon co-loading with RuO2 [3]. The band gaps have been noted to be highly dependent of the Zn content [4]. The high temperature nitridation process gives very limited control over the final Zn content and this coupled with a significant surface area loss due to sintering (< 8 m2/g) makes them a less attractive alternative to TiO2 though they have the ability to be active in visible light (band gaps of 2.6 to 2.8 eV).

Our goal is to improve current or develop new synthesis that optimize the Zn/Ga ratio closer to 0.5 without subsequent surface area or Zn loss from the precursor. We are currently following a multi-pronged approach to tackle these two issues; Zn-O bond strength is dependent on the precursor employed and is lower for ZnO compared with ZnGa2O4 [5]. Employing this we were able to synthesize ZGONs with relatively low band gaps (2.35 eV) and adequate surface areas (36 m2/g). Interestingly the materials no longer tend to produce a wurzite (tetrahedral Ga) crystal structure as reported earlier3 but tend to be a spinel (octahedral Ga) though it has essentially the same Zn and Ga content according to SEM-EDAX. We are currently investigating the spinel ZGON structure using XRD, EXAFS, Ga71 NMR, XPS, and Raman spectroscopic techniques. Increasing surface areas can also be achieved by the preparation of photocatalysts with mesoporous silica (SBA 15). Zinc gallate phase was dispersed in SBA15 and finally nitrided to yield a total surface area of 515 m2/g. Thus, the above improvements shown in the properties (band gaps, surface areas) can lead to better photocatalysts for water splitting to produce hydrogen. These methods (among others) are being investigated to synthesize catalysts which can replace TiO2 with one that can be effectively be utilized the solar spectrum eliminating the need to use energy intensive UV light sources.

1. Fujishama, A.; Zhang, X., Chimie 2005, 8, 750-760

2. Maeda, K.; Domen, K., Journal of Physical Chemistry C 2007, 111, (22), 7851-7861

3. Maeda, K.; Teramura, K.; Lu, D.; Takata, T.; Saito, N.; Inoue, Y.; Domen, K., Nature 2006, 440, (7082), 295-295

4. Jensen, L. L.; Muckerman, J. T.; Newton, M. D., Journal of Physical Chemistry C 2008, 112, (9), 3439-3446

5. Chen, H.; Wen, W.; Wanq, Q.; Hanson, J. C.; Muckerman , J. T.; Fujita, E.; Frenkel, A. I.; Rodriquez, J. A.; Journal of Physical Chemistry C 2009, 113, (9), 3650-3659

See more of this Session: Nanomaterials for Hydrogen Production and Fuel Cells I
See more of this Group/Topical: Topical 5: Nanomaterials for Energy Applications

Solar Hydrogen Production Using Ferrite-Based Water Splitting Cycles On Al_2O_3 Substrates

Jonathan R. Scheffe, Department of Chemical and Biological Engineering, University of Colorado, Boulder, CO

Jianhua Li, Department of Chemical and Biological Engineering, University of Colorado, Boulder, CO

Alan W. Weimer, Department of Chemical and Biological Engineering, University of Colorado, Boulder, CO

We are reporting the results of an alternative ferrite cycle in which the ferrite ($CoFe_2O_4$) is deposited on a porous Al_2O_3 support via atomic layer deposition (ALD). Rather than a traditional cycle, in which the ferrite is deposited on an inert support or is unsupported, the ferrite reacts with Al_2O_3 during thermal reduction to form hercynite ($FeAl_2O_4$). This is advantages because the reaction between the ferrite and Al_2O_3 is favored at lower temperatures than the reduction of Fe^{3+} to Fe^{2+} on an inert support. Subsequent water oxidation is then achievable at 1000 °C, similar to other ferrite cycles. This redox cycle is shown in the two step reaction below:

$CoFe_2O_4 + Al_2O_3$ + solar thermal energy => $CoO + 2FeAl_2O_4 + 0.5\ O_2$ (1)

$CoO + 2FeAl_2O_4 + H_2O \Rightarrow CoFe_2O_4 + Al_2O_3 + 2H_2$ (2)

In the first step, solar thermal energy is used to reduce the ferrite, and oxygen is evolved. In the second step, the reduced material is reacted with water to generate hydrogen and is reoxidized. Therefore, the only inputs are solar energy and water, and the only outputs are H_2 and O_2.

Thermal reduction was attempted at temperatures ranging from 1200 °C to 1500 °C, and results indicate that reduction occurs at lower temperatures than traditional ferrite cycles. It was observed that hercynite ($FeAl_2O_4$) was formed after reduction, due to a reaction between FeO and the Al_2O_3 substrate.

Subsequent water oxidation was achievable at 1000 °C and the reaction was shown to be repeatable over 8 cycles. Experimental results are compared to thermodynamic modeling performed using FactSage and results agree well with thermodynamic predictions, as seen in figure 1.

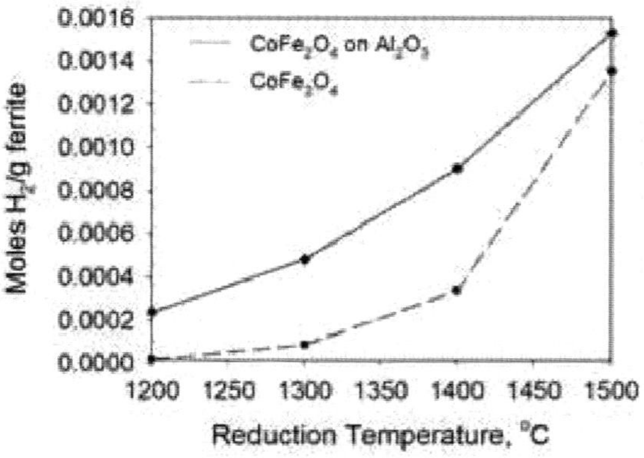

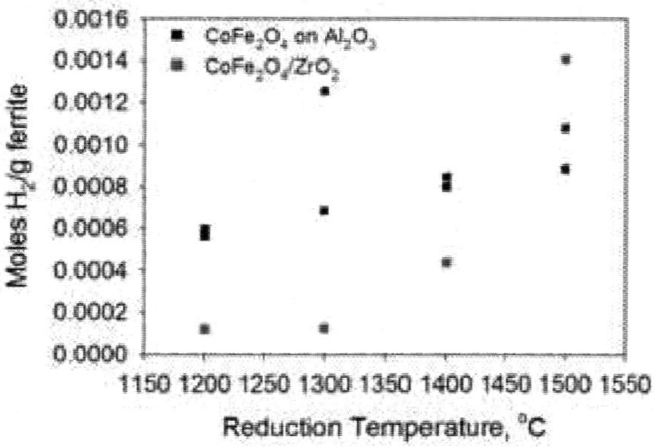

Figure 1. Left) Thermodynamic modeling results predict more H_2 generation for $CoFe_2O_4$ on Al_2O_3 over reduction temperatures from 1200 to 1500 °C. Right) Experimental results predict more H_2 generation for $CoFe_2O_4$ on Al_2O_3 over reduction temperatures from 1200 to 1400 °C

H_2 conversions varied between 18% and 25 % when cycling was performed at 1200 °C reduction and 1000 °C water oxidation. Crystallinity changes were measured before and after reduction using powder x-ray diffractometry, and conversions were calculated via in situ mass spectrometry. Film morphology and composition were measured via high resolution transmission electron microscopy and induced coupled plasma-atomic emission spectroscopy, respectively.

See more of this Session: Nanomaterials for Hydrogen Production and Fuel Cells I
See more of this Group/Topical: Topical 5: Nanomaterials for Energy Applications

Gold-TiO_2 Nanostructured Photoanodes for Photoelectrochemical Cells

Peter H. Aurora, Mechanical Engineering, University of Michigan, Ann Arbor, MI

Chang Hwan Kim, Chemical Engineering, University of Michigan, Ann Arbor, MI

Levi T Thompson, Chemical Engineering, University of Michigan, Ann Arbor, MI

Photoelectrochemical cells (PEC) combine a photovoltaic device for light harvesting and an electrolyzer for water splitting into a single system. These cells consist of a photoanode for water oxidation and oxygen evolution, a cathode where hydrogen is evolved, and an electrolyte. A major obstacle for the use of PEC cells is low photoanode efficiency. By some accounts, the rate of water oxidation has to be increased by more than an order of magnitude to keep pace with the production of electrons and holes. In addition, recombination losses can limit performance. For example, when particulate catalysts are used in the photoanode only about 5% of the generated carriers are available for the reaction. We explored two strategies to improve the photoanode performance: producing semiconducting oxides in the form of nanotubes and incorporating nanoscale gold onto their surfaces.

Highly ordered TiO_2 nanotube (TiNT) arrays were fabricated using an anodization process. Gold nanoparticles were supported onto the TiNTs using a modified deposition precipitation method. The pH, aging time and Au precursor concentration were manipulated to increase the Au dispersion and loading on the nanotubes. Cyclic voltammetry was used to measure the total nanotube and Au electrochemical surface areas, and to study the location of traps and active states in the photocatalyst. The performance of photoanodes was evaluated in a 3-electrode cell with a 1.0 M KOH electrolyte using a solar simulator (1.5 AM) and potentiostat/galvanostat. Scanning electron microscopy (SEM), transmission electron microscopy (TEM), and X-ray diffraction were used to characterize key properties of the materials, and optical absorption measurements were used to determine their bandgaps.

Titania nanotubes with a range of morphologies have been produced and the resulting electrodes showed improved efficiencies perhaps as a consequence of decreased recombination losses by enhancing capture of the light and better separating and transporting charge carriers. The incorporation of Au nanoparticles resulted in a slight reduction of the bandgap, which has been attributed to the existence of impurity levels between the band edges of the oxide. Introduction of the gold nanoparticles resulted in a significant improvement in the electrocatalytic properties. The photocurrent normalized by the Au active area increased dramatically as the size of Au particles decreased below 5 nm. This increase may have been a consequence of more effective hole transfer from TiO_2 to Au, thereby enhancing the H_2O oxidation rates. Experiments are under way to determine the structural and compositional character of active sites involved in H_2O oxidation.

Longer nanotubes provided higher photocurrents compared to TiO_2 powders and short nanotubes. This was attributed to better separation of the charge carriers, as well as enhanced capture of the light. The incorporation of Au nanoparticles improved rates for water oxidation to produce hydrogen. Catalysts with smaller Au nanoparticles were significantly more active than those containing larger Au particles.

See more of this Session: Nanomaterials for Hydrogen Production and Fuel Cells I
See more of this Group/Topical: Topical 5: Nanomaterials for Energy Applications

Low Temperature H2 Production From Biomass Over Transition Metal Oxide Nanoparticles

Zhong He, Chemical, Biological and Pharmaceutical Engineering, New Jersey Institute of Technology, Newark, NJ

Xianqin Wang, Chemical, Biological and Pharmaceutical Engineering, New Jersey Institute of Technology, Newark, NJ

Nowadays, over 90% H2 is still produced from steam reforming of natural gas. This process requires high temperature operation. Moreover, the natural gas is mostly provided from fossil fuel resources. In order to minimize the environmental impact, hydrogen produced from biomass resource has been a critical issue to the Department of Energy (DOE) in USA and abroad. The synthesis of functional nanomaterials offers a unique opportunity to make the breakthrough discoveries and truly revolutionary developments to meet these challenges. The work presented here is to produce H2 from biomass by using Pt clusters supported on transition metal oxide nanoparticles. The results have shown that the strong interaction between transition metal oxide nanoparticles and Pt nanoparticles increased the hydrogen production and lowered the reaction temperature significantly.

See more of this Session: Nanomaterials for Hydrogen Production and Fuel Cells I
See more of this Group/Topical: Topical 5: Nanomaterials for Energy Applications

A-Fe2O3 Nanowire Array Based Electrodes for Self-Driven Photoelectrochemical Cells

Boris D. Chernomordik, Department of Chemical Engineering, University of Louisville, Louisville, KY

Uros Cvelbar, Plasma Laboratory F4, Jozef Stefan Institute, Ljubljana, Slovenia

Jacek B. Jasinski, Institute of Advanced Materials and Renewable Energy, University of Louisville, Louisville, KY

Miran Mozetic, Department for Surface Engineering, Jozef Stefan Institute, Ljubljana, Slovenia

Mahendra K. Sunkara, Department of Chemical Engineering, University of Louisville, Louisville, KY

Iron oxide, Fe_2O_3, is a promising material for water splitting reaction using solar energy.[1-2] Even the recent efforts, however, using Fe_2O_3 thin film materials reported low efficiencies due to poor carrier transport within these films.[3-4] In this regard, we have been working on a-Fe_2O_3 nanowire arrays as possible thin film materials systems for improved carrier mobility during photoelectrochemical water splitting reaction.

Recently, we developed a novel method to synthesize arrays of α-Fe_2O_3 nanowires, which are single crystal and have highly ordered oxygen vacancy plane.[5-6] The nanowire arrays are directly synthesized on metallic substrates using low-temperature RF plasma oxidation without the use of any foreign metal catalysts. The nanowire growth is spontaneous and the process lasts only seconds. This method could be scaled up for manufacturing of large area nanowire arrays on metallic foils. As one-dimensional nanostructures, these nanowires offer many other benefits to PEC electrolysis, such as high surface area, reduced charge carrier diffusion distance, and a preferential direction for charge diffusion. In this presentation, we will provide both synthesis and their photoelectrochemical properties of a-Fe_2O_3 nanowire arrays grown on Fe foils.

References

1. K.L. Hardee, A.J. Bard, Semiconducting electrodes. 5. Application of Chemically Vapor Deposited Iron Oxide Films to Photosensitized Electrolysis", J. of the Electrochemical Society 1976, 123, (7), 1024-1026.

2. J.H. Kennedy, M. Anderman, Photoelectrolysis of Water at alpha- Fe_2O_3 electrodes in Acidic Solution", J. of the Electrochemical Society 1983, 130, (4),

3. T. Lindgren, H.L. Wang, N. Beermann, L. Vayssieres, A. Hagfeldt, S.E. Lindquist, Aqueous photoelectrochemistry of hematite nanorod array, "Solar Energy Materials and Solar Cells", 2002, 71, (2), 231-243.

4. W.B. Ingler, S.U.M. Khan, "A self-driven p/n-Fe2O3 tandem photoelectrochemical cell for water splitting", Electrochemical and Solid State Letters 2006, 9, (4), G144-G146.

5. U. Cvelbar, Z. Chen, M.K. Sunkara and M. Mozetic, "Spontaneous growth of superstructure α-Fe2O3 nanowire and nanobelt arrays in reactive oxygen plasma", Revision Submitted, Small (2008).

6. Z. Chen, U. Cvelbar, M. Mozetic, J. He and M.K. Sunkara, "Long-Range Ordering of Oxygen-Vacancy Planes in r-Fe2O3 Nanowires and Nanobelts", Chemistry of Materials, 20 (9), 3224 - 3228 (2008).

See more of this Session: Nanomaterials for Hydrogen Production and Fuel Cells I
See more of this Group/Topical: Topical 5: Nanomaterials for Energy Applications

Nano-Scale Atomic Layer Deposited Films for Water Splitting Cycles

Victoria J. Aston, Department of Chemical and Biological Engineering, University of Colorado, Boulder, CO

Christopher Perkins, Sundrop Fuels, Louisville, CO

Alan W. Weimer, Department of Chemical and Biological Engineering, University of Colorado, Boulder, CO

Iron oxide (Fe_3O_4) can be reduced to iron metal using synthesis gas (CO, H_2, CO_2) at moderate temperatures (700 - 1200°C). Iron metal is then used to split water at lower temperatures (300 - 600°C), producing a stream of pure, uncontaminated hydrogen and regenerating the Fe_3O_4 for syngas reduction again. However, both the reduction and oxidation reactions are controlled by diffusion through the reacted layer, making the reaction slow when bulk iron is used. The higher temperature reduction step also results in sintering of the powder, decreasing the surface area for diffusion and greatly slowing the reaction rate. This limits the number of reduction and oxidation cycles that bulk iron can undergo. By coating high surface area substrates with nano-thin films of iron using Atomic Layer Deposition (ALD), the reaction can be completed more quickly than with bulk iron powder for the same mass of iron. The thin films and high surface area substrates also mitigate the effects of sintering. In this study, the kinetics of the syngas reduction and water oxidation of these iron thin films on different substrate materials (SiO_2, Al_2O_3, and ZrO_2) have been studied and compared with the kinetics of the reactions when using bulk iron powder.

See more of this Session: Nanomaterials for Hydrogen Production and Fuel Cells I
See more of this Group/Topical: Topical 5: Nanomaterials for Energy Applications

Dye-Sensitized Solar Cells Based On Ordered TiO2 Nanotube Arrays On Transparent Conductive Oxide

Chengkun Xu, Chemical and Petroleum Engineering, University of Pittsburgh, Pittsburgh, PA

Paul Shin, Chemical and Petroleum Engineering, University of Pittsburgh, Pittsburgh, PA

Liangliang Cao, Chemical and Petroleum Engineering, University of Pittsburgh, Pittsburgh, PA

Jiamin Wu, Chemical and Petroleum Engineering, University of Pittsburgh, Pittsburgh, PA

Di Gao, Chemical and Petroleum Engineering, University of Pittsburgh, Pittsburgh, PA

Dye-sensitized solar cells (DSCs) are promising for low-cost solar electricity production. Anodes of current DSCs use mesoporous films of TiO2 nanoparticles. Although such films provide a high surface area for dye adsorption, their disordered pore structure and short electron diffusion length have limited further improvement of DSCs. ZnO nanowire-based anodes have shown promise to overcome these constraints, but their performance is limited by the intrinsic material properties of ZnO. Here, we present a DSC based on vertically aligned TiO2 nanotube arrays synthesized directly on transparent conductive oxides. It is found that the TiO2 nanotube-based DSCs significantly outperform ZnO nanowire-based ones, and the electron lifetime in TiO2 nanotubes is more than 10 times longer than those in ZnO nanowires and TiO2 nanoparticles, which leads to significantly improved electron-collecting efficiency. This result opens up numerous opportunities to further improve DSCs, for example, by employing solid-state electrolytes and redox mediators with faster kinetics.

See more of this Session: Nanomaterials for Photovoltaics I
See more of this Group/Topical: Topical 5: Nanomaterials for Energy Applications

Reaction Mechanism for the Formation of CuInSe2 Nanocrystals for Low-Cost Solar Cells

Mahaprasad Kar, School of Chemical Engineering and Energy Center at Discovery Park, Purdue University, West Lafayette, IN

Hugh W. Hillhouse, School of Chemical Engineering and Energy Center at Discovery Park, Purdue University, West Lafayette, IN

Rakesh Agrawal, School of Chemical Engineering and Energy Center at Discovery Park, Purdue University, West Lafayette, IN

The development of suitable colloidal nanocrystal inks are a key step in the development of low-cost solar cells since they enable the use of fast and inexpensive coating processes such as spray coating and roll coating to form a thin film photo-absorbing layer. Guo et al. recently reported the solution-phase synthesis of chalcopyrite structured CuInSe2 (CISe) nanocrystals from elemental and halide reagents via a simple batch reaction in a single component coordinating solvent [1]. The use of these nanocrystals for solar cells has been demonstrated by fabricating devices, with the first solar cells having an efficiency of 2.8%. Insights into the reaction mechanism for the formation of the nanocrystals are a key step to gain a better control over the electronic properties of the ternary semiconductor. A reaction mechanism for the formation of CISe nanocrystals is proposed using powder x-ray diffraction, electron microscopy, compositional analysis, FT-IR spectroscopy, mass spectrometry and thermodynamic calculations. The formation of CISe nanocrystals from metal-chloride precursors and elemental selenium in oleylamine is preceded by the formation of copper selenide (CuSe) and indium selenide (InSe) as the solid-phase intermediates. And the reaction takes place via the chlorination of the methylene groups in oleylamine with hydrochloric acid being the other liquid-phase by-product. However, it has been observed that the reaction of CuSe with indium chloride (InCl3) dissolved in oleylamine leads to the formation of the ordered chalcopyrite structure of CISe. In support of the proposed reaction mechanism and contrary to previous reports in the literature, chalcopyrite CISe nanocrystals have also been synthesized in solvents without any metal-coordinating groups.

References:

1. Guo, Q., S.J. Kim, M. Kar, W.N. Shafarman, R.W. Birkmire, E.A. Stach, R. Agrawal, and H.W. Hillhouse, Development of CuInSe2 nanocrystal and nanoring inks for low-cost solar cells. Nano Letters, 2008. 8(9): p. 2982-2987.

See more of this Session: Nanomaterials for Photovoltaics I
See more of this Group/Topical: Topical 5: Nanomaterials for Energy Applications

Synthesis of Colloidal Cu(In,Ga)Se_2 (CIGS) Nanocrystal "Inks" and Their Implementation Into Photovoltaic Devices

Vahid A. Akhavan, Chemical Engineering, University of Texas at Austin, Austin, TX

Brian W. Goodfellow, Chemical Engineering, University of Texas at Austin, Austin, TX

Matthew G. Panthani, Chemical Engineering, University of Texas at Austin, Austin, TX

Brian A Korgel, Chemical Engineering, University of Texas at Austin, Austin, TX

Copper indium gallium selenide (CIGS) based solar cells show great promise for widespread use in single junction thin film devices. Large scale production of the CIGS absorber layer, however, is hindered by high cost and poor stoichiometric control associated with the coevaporation of tertiary and quaternary systems in high vacuum. We have devised an inexpensive, solution-based deposition of CIGS under ambient conditions with predetermined stoichiometry. CIGS nanocrystals approximately 10 nm in diameter were synthesized by arrested precipitation and dispersed in organic solvents as "inks". We have demonstrated room-temperature deposition of these inks by dip-coating, drop-casting, ink-jet printing, or airbrush spraying to yield device quality films. The ability to deposit CIGS layers under ambient conditions permits alternate device architectures not possible with evaporated CIGS. We have demonstrated functional conventional and inverted device structures employing a variety of back contact materials. An inverted structure removes the need for the CdS buffer layer, making the device ecologically sound. Additionally, we have fabricated functional devices on flexible substrates to demonstrate the ability to integrate this technology into a roll-to-roll production scheme.

Photovoltaic devices fabricated by nanocrystal deposition under ambient conditions have exhibited power conversion efficiencies of 1.5% under AM1.5 illumination. The device performance to date has been limited by the high grain boundary density in the CIGS nanocrystal films that leads to high recombination currents. Films with much larger crystalline CIGS domains and lower grain boundary density can be created by annealing the nanocrystal films at high temperature (500°C) under a selenium atmosphere. Grain sizes up to 10 μ m can be produced and the devices exhibit much reduced surface recombination and higher efficiencies. With the appropriate processing steps, nanocrystal-based PV devices are capable of reaching commercially viable device efficiencies.

See more of this Session: Nanomaterials for Photovoltaics I
See more of this Group/Topical: Topical 5: Nanomaterials for Energy Applications

Nanomanufacturing of Ordered Plasmonic Nanostructures: Effect of Metal-Substrate Interactions

Lucas Apollo Lane, Energy, Environmental, and Chemical Engineering, Washington University, St. Louis, MO

R. Sureshkumar, Department of Energy, Environmental and Chemical Engineering and the Center for Materials Innovation, Washington University, Saint Louis, MO

Ordered metal nanoparticles have gained great importance in recent research activities due to their potential applications in nonlinear optics and nanophotonics below the diffraction limit. These enhanced optical properties arise from the interactions between the photons and the resonant localized surface plasmons, i.e., the collective oscillation of the free electron gas density at the metal/dielectric interface. An important example of one of the many applications that these nanoparticle arrays is that they can act as broadband antennas for trapping sunlight in thin film solar cells. These thin film solar cells have the potential to reduce the cost of photovoltaics while increasing light absorption leading to an increase in efficiency. A route to make such ordered structures, while still being practical and cost effective, can be through instabilities of an ultra thin (1-10 nm) fluid film leading to spatial patterns with length and time scales that are dependent on properties such as interfacial tension, contact angle with the substrate, and long range dispersion forces such as the van der Waals interaction [Favazza et al. Nanotechnology 17 (2006) 4229-4234, Trice et. al. Physical Review Letters 101, 017802 (2008)]. In such a process a metallic film is deposited on a substrate (e.g. silica) by physical deposition and subsequently the metal-substrate bilayer system is irradiated with a nanosecond or femtosecond pulsed laser. The laser fluence is selected to be above the melting threshold of the metal. During the pulse, the liquid film undergoes dewetting, resulting in the formation of a variety of spatially ordered nanostructures with a broad range of plasmon responses. In this work we study the added effects of a diffusion potential on the stability of the thin film. This potential represents the chemical affinity of the metal to the substrate such as the one that exists between metals and Si. Significant increases of the diffusion potential through the film have the ability to damp perturbations caused by forces such as the van der Waals interactions and increase the length scales between the particles. A new parameter is identified to quantify the effect of film-substrate interactions on the dampening of the perturbations. Implications to plasmon based solar energy harvesting applications will be discussed.

See more of this Session: Nanomaterials for Photovoltaics I
See more of this Group/Topical: Topical 5: Nanomaterials for Energy Applications

Electrochemical Synthesis and Characterization of Organic/Inorganic Hybrid Nanostructures for Photovolatics

Chong Hyun Chang, Chemical and Environmental Engineering, University of California-Riverside, Riverside, CA

Jae-Hong Lim, Chemical and Environmental Engineering, University of California-Riverside, Riverside, CA

Nosang V. Myung, Chemical and Environmental Engineering, University of California-Riverside, Riverside, CA

Hybrid nanostructures based photovoltaics have gained interest in the recent years because of the ability to enhance the energy conversion efficiency in cost-effective matter. In this work, conducting polymers (i.e. PEDOT:PSS, P3HT) and II-VI semiconducting (i.e. CdTe) nanoparticles coated carbon nanotube (CNT) networks were electrochemically synthesized. The coating thickness, size and number of nanoparticles, and composition were preciously controlled by adjusting electrodeposition conditions including applied charge and potential. Their electrical and optoelectronic properties were systematically investigated to determine optimum conditions. The field effect transistor characteristics and photocurrent response of CdTe nanoparticles decorated SWNTs were strongly depended on the content of semiconducting SWNTs compared to metallic SWNTs and number of CdTe nanoparticles. Similar to CdTe, electronic and optoelectronic properties of PEDOT:PSS coated CNTs were characterized and optimized to enhance the properties.

See more of this Session: Nanomaterials for Photovoltaics I
See more of this Group/Topical: Topical 5: Nanomaterials for Energy Applications

Nanoscale Optical, Electrical, and Chemical Interrogation of Thiophene-Based Polymer Solar Cells

Isaac Riisness, Chemical Engineering, UCSB, Santa Barbara, CA

Raphael Ramos, Chemical Engineering, UCSB, Santa Barbara, CA

Chris Carach, Chemical Engineering, UCSB, Santa Barbara, CA

Michael Gordon, Chemical Engineering, University of California Santa Barbara, Santa Barbara, CA

Recent advances in the synthesis, functionalization, and processing of conjugated polymers have brought organic semiconductors to the forefront of photovoltaics research. In particular, bulk heterojunctions (BHJs) created by blending a photoactive donor polymer with a fullerene-based electron acceptor have the shown the highest efficiency and most promise. The BHJ device structure is appealing because exciton harvesting is local (i.e., donor and acceptor domains are smaller than typical exciton diffusion lengths) and both the donor polymer and acceptor can be deposited from a common solvent, allowing flexible and large area devices to be fabricated using low-cost solution processing techniques. Despite these promising attributes, the BHJ faces several key challenges such as low efficiency, oxidation damage to the donor polymer backbone leading to charge trapping and recombination centers, and poorly understood donor-acceptor interfaces. In fact, our ability to synthesize new materials and device structures has far outstretched our ability to characterize and understand them at length scales relevant to their operation.

To this end, our work focuses on confocal and near-field optical and electrical interrogation of conjugated polymers and polymer composites to understand how nanoscale morphology, molecular alignment, and chemical structure affect charge transport processes. In this talk, we show that wavelength-resolved confocal, near-field Raman, and photoluminescence (PL) imaging of fullerene-doped (PCBM) polythiophene, when combined with electrical AFM techniques, are highly descriptive means to study donor and acceptor mixing, oxidation effects, and charge transport. For example, spatially-correlated confocal PL and Raman imaging can easily differentiate between regions of low PL due to good blending or simply a lack of P3HT donor. We will also present nanoscale electrical and chemical maps of P3HT/PCBM blends to show that near-field vibrational spectroscopy, in a scanning probe microscopy platform, is a powerful technique to study the nanoscale morphology in BHJ films.

See more of this Session: Nanomaterials for Photovoltaics I
See more of this Group/Topical: Topical 5: Nanomaterials for Energy Applications

Photovoltaic Electrochemical Optical Rectennas of Carbon Nanotubes with Self-Assembled Monolayers

Juan G. Duque, Department of Chemical and Biomolecular Engineering, Rice University, Houston, TX

Matteo Pasquali, Department of Chemical and Biomolecular Engineering, Rice University, Houston, TX

Howard K. Schmidt, Chemical and Biomolecular Engineering, Houston, TX

Carbon nanotubes (CNT) and CNT-gold hybrid nanostructures are shown to produce rectified photocurrents without pre-bias when illuminated in the presence of aqueous electrolytes containing anionic surfactants. Graphite nanoplatelets, multi-wall carbon nanotube mats, single-wall carbon nanotube (SWNT) arrays and carbon black-loaded polymer all display this novel photovoltaic effect, and share a unique isolated photocurrent peak at ~300 nm. This feature apparently originates with their shared π-plasmon absorption. Rectified photocurrents are obtained only in the presence of anionic surfactants known to form self-assembled monolayers on graphitic surfaces. Carbon nanotube electrodes also produce minor photocurrents in the near infra-red. When coated with 5 nm of gold, single-walled carbon nanotube arrays generate similar photocurrents when treated with thiol-terminated anionic surfactant molecules. The Au-SWNT composite nanostructure displays a broad photocurrent spectrum spanning the entire visible region, along with a pronounced peak at ~400 nm. These features appear to arise from photoabsorption by gold nanorod plasmon hybridized with SWNT π-plasmons. A model explaining these results based on hot carriers traversing a rectifying tunnel barrier is presented and discussed. This mechanism is consonant with the concept of an optical rectifying antenna (rectenna), and could enable development of high efficiency photodetectors, photovoltaics and photochemical converters.

See more of this Session: Nanomaterials for Photovoltaics II
See more of this Group/Topical: Topical 5: Nanomaterials for Energy Applications

CdS Sensitized Nanostructured TiO2 and ZnO Solar Cells

Jie Pan, Paper and Chemical Engineering, Miami University, Oxford, OH

Lei L. Kerr,Paper and Chemical Engineering, Miami University, Oxford, OH

A disadvantage of using organic dye in current dye sensitized solar cell technology is that it can rapidly deteriorate under the illumination of UV light. This instability can possibly be overcome by using quantum dots or nano-sized particles of CdS as an alternative sensitizer. The effects of post annealing treatment were investigated on the CdS sensitized ZnO and TiO2 nanocrystalline solar cells. It was found that the annealing will increase the cell efficiency on ZnO and TiO2 with porous structures. A significant increase of cell efficiency from 0.54% to 1.07% was observed for CdS sensitized porous ZnO solar cells after annealing and 0.52% to 1.29% for CdS sensitized electrospun porous TiO2 electrode. The increase is attributed to the film morphology improvement and the absorption band edge shift of CdS. The annealing does not increase the efficiency for solar cells made from CdS sensitized TiO2 without porous structure.

See more of this Session: Nanomaterials for Photovoltaics II
See more of this Group/Topical: Topical 5: Nanomaterials for Energy Applications

Electron Transfer and Transport Measured in Nanotube and Nanoparticle Photoanodes by Ultrafast THz Spectroscopy

Christiaan Richter, Department of Chemistry, Yale University, New Haven, CT

Charles A. Schmuttenmaer, Department of Chemistry, Yale University, New Haven, CT

Dye sensitized solar cells (DSSC's) make use of TiO2 nanoparticle films as a high surface area support for light harvesting dye molecules. It has been proposed that TiO2 nanotubes could be an alternative high surface area dye support that may have superior electron transport properties to random nanoparticle networks. However, macroscopic photocurrent measurements of fully functional solar cells suggest electron transport rates in titania nanotube and nanoparticle films to be more or less comparable. However, photovoltage measurements reveal that nanotubes have significantly lower energy loss due to electron recombination. Through its interaction with free carriers far-infrared (FIR) light can be used as a non-contact probe for observing free carriers in nanostructures directly. Using ultra-short FIR laser pulses the free electron mobility and population can be tracked with a temporal resolution on the order of 10 to 100 femtoseconds. These considerations render time-resolved terahertz spectroscopy (TRTS) an ideal probe with which to confirm, and obtain a fundamental understanding of, the macroscopic electron transport measurements mentioned above. Charge injection into TiO2 nanotubes and nanoparticles was monitored on sub-picosecond to nanosecond times scales by time-resolved terahertz spectroscopy (TRTS). Dye molecules injected electrons into both the semiconductor nanoparticles and nanotubes after photoexcitation with a 400 nm pump pulse, causing an increase in the semiconductor's electron density that can be detected and monitored using a THz probe pulse. To study free electron mobility and dynamics within nanotubes and particles the time-resolved frequency-dependent photoconductivity was measured. The nanotube and nanoparticle photoconductivity was obtained by measuring the transmitted far-infrared electric field. Electron mobility, electron scattering times, and electron lifetimes within nanotubes and particles were extracted from the data. The variation of these electron transport properties was studies as a function of nanostructure (nanotubes vs. nanoparticles), nanotube morphology (wall thickness, diameter etc.) and crystallinity (annealing temperature and time). The implications of these findings for understanding and optimizing the functioning of DSSC's and electrical transport in disordered media will be discussed.

See more of this Session: Nanomaterials for Photovoltaics II
See more of this Group/Topical: Topical 5: Nanomaterials for Energy Applications

Photosystem I-Based Biohybrid Systems for Photoelectrochemical Catalysis

Peter N. Ciesielski, Department of Chemical and Biomolecular Engineering, Vanderbilt University, Nashville, TN

Frederick Hijazi, Chemical and Biomolecular Engineering, Vanderbilt University, Nashville, TN

Amanda Scott, Chemical Engineering, Vanderbilt University, Nashville, TN

Christopher J. Faulkner, Department of Chemical and Biomolecular Engineering, Vanderbilt University, Nashville, TN

David Cliffel, Chemistry, Vanderbilt University, Nashville, TN

G. Kane Jennings, Department of Chemical and Biomolecular Engineering, Vanderbilt University, Nashville, TN

Photosynthesis is the process by which Nature coordinates a tandem of protein complexes of impressive complexity that function to harness staggering amounts of solar energy on a global scale. Advances in biochemistry and nanotechnology have provided tools to isolate and manipulate the individual components of this process, thus opening the door to a new class of highly functional and vastly abundant biological resources. Here we show how one of these components, Photosystem I (PSI), is incorporated into non-biological systems to photocatalyze electrochemical reactions that produce an electrical current. We demonstrate that this process can be enhanced by moving from flat electrodes to high surface area, nanoporous electrodes which accommodate more PSI complexes and thus produce larger photocurrents. We also describe the fabrication of stand-alone photoelectrochemical cells in which the functions of light absorption and charge separation are performed by PSI complexes that are assembled into a dense multilayer on the cell's cathode, and we report the performance of these cells under various intensities of white light.

See more of this Session: Nanomaterials for Photovoltaics II
See more of this Group/Topical: Topical 5: Nanomaterials for Energy Applications

Self-Assembled Biomimetic Antireflection Coatings for High Efficiency Photovoltaics

Wei-Lun Min, Chemical Engineering, University of Florida, Gainesville, FL

Chih-Hung Sun, Chemical Engineering, University of Florida, Gainesville, FL

Nicholas Linn, Chemical Engineering, University of Florida, Gainesville, FL

Bin Jiang, Mathematics and Statistics, Portland State University, Portland, OR

Peng Jiang, Chemical Engineering, University of Florida, Gainesville, FL

Millions of years before we began to generate functional nanostructures, biological systems were using nanometer-scale architectures to produce unique functionalities. Some nocturnal moths use hexagonal arrays of subwavelength nipples as antireflection coatings (ARCs) to reduce reflection from their compound eyes. Similar periodic arrays of pillars have also been observed on the wings of cicada to render superhydrophobic surfaces for self-cleaning functionality. Inspired by these natural nanostructures, we have developed a simple and scalable templating technique for fabricating self-cleaning, broadband ARCs on a large variety of technologically important substrates (e.g., crystalline Si, GaAs, GaSb, glass, and polymer). The technique is based on a spin-coating platform that combines the simplicity and cost benefits of bottom-up colloidal self-assembly with the scalability and compatibility of standard top-down microfabrication. The resulting subwavelength-structured ARCs exhibit superior broadband antireflection properties than traditional multilayer dielectric ARCs and are promising for applications ranging from highly efficient solar cells and photodetectors to flat-panel displays.

See more of this Session: Nanomaterials for Photovoltaics II
See more of this Group/Topical: Topical 5: Nanomaterials for Energy Applications

Fuel Cell Electrocatalysts Obtained From Bimodal Nano-Porous Templates

Svitlana Pylypenko, University of New Mexico, Albuquerque, NM

Tim S. Olson, University of New Mexico Chemical Engineering, Albuquerque, NM

Dimiter Petsev, Chemical and Nuclear Engineering, University of New Mexico, Albuquerque, NM

Plamen Atanassov, Department of Chemical Engineering, University of New Mexico, Albuquerque, NM

The development of hierarchically-structured electrocatalysts and their supports can effectively address some of the current limitations of fuel cells. It is highly desirable to decrease Pt loading, while increase the obtainable power densities and improve durability of fuel cells. One way to increase the current density is through designing of the electrocatalyst not only with high surface area but also with high amount of accessible three-phase sites. Porous structures can effectively minimize transport limitations, thus increasing the accessibility of the active sites by gas and electrolyte phases. The durability requirements of the fuel cells can be met through improvement of the corrosion stability of the current supports and design of novel corrosion resistant supports. The procedure discussed here allows to create materials that will address both, mass-transport and durability limitations of the fuel cells.

Hierarchical bimodal porous silica templates were obtained through oil/water/surfactant microemulsion templating. The first type of pores is in the range of 10-40 nm and is determined by dimensions of microemulsion droplets. The second type of pores is around 5 nm and is determined by micellar dimensions.

Silica was used as a template to make a range of materials. First, carbon supported platinum electrocatalysts will be discussed. The role of the small nanopores here is to lock Pt particle size in the range of 3-5 nm. Next, non-noble electrocatalysts based on pyridinic and cobalt precursors will be discussed. Here, small nanopores of the silica template play role of sites for creation of Me-Nx centers. Large nanopores are channels necessary for complete infiltration of the template with precursor materials and formation of the catalyst and/or catalyst support backbone. Synthesis conditions, such as amount of precursors, temperature of pyrolysis and conditions of removal of the silica template play an important role in formation of porous structure of electrocatalyst. Ability to control these conditions enables to create electrocatalytic materials with better transport properties.

See more of this Session: Nanomaterials for Energy Production and Fuel Cells II
See more of this Group/Topical: Topical 5: Nanomaterials for Energy Applications

Synthesis of Pt-Modified Tungsten Monocarbide and Its Activity towards the Hydrogen Evolution Reaction

Daniel V. Esposito, Chemical Engineering, University of Delaware, Newark, DE

Kevin D. Dobson, Institute of Energy Conversion, University of Delaware, Newark, DE

Brian E. McCandless, Institute of Energy Conversion, University of Delaware, Newark, DE

Robert W. Birkmire, Institute of Energy Conversion, University of Delaware, Newark, DE

Jingguang G. Chen, Chemical Engineering, University of Delaware, Newark, DE

Hydrogen is a promising energy carrier for the future because it represents a means for carbon-free energy storage and can be produced from biomass or water using renewable energy. When produced from water in an electrolyzer or photoelectrochemical cell (PEC), $H_2(g)$ is evolved at the cathode via the hydrogen evolution reaction (HER) while water is split at the anode to produce $O_2(g)$. For electrolysis in acidic solutions, the best HER catalysts are precious metals such as Pt, Rh, and Pd [1], but their prohibitively high costs and limited supplies are major hurdles for the widespread production of H_2 by this means. One approach for overcoming this challenge is to disperse the active precious metal catalyst in sub-monolayer or monolayer coverages on a low-cost, stable, supporting catalyst.

In this work, we investigate the performance of tungsten monocarbide (WC), modified by sub-monolayer amounts of Pt, as a HER catalyst. WC is an attractive material for this purpose because it has shown an appreciable activity for the HER [2], is known for its electrocatalytic stability in acidic solutions [3], and has demonstrated synergistic effects when combined with Pt as an electrocatalyst [4]. In recent years, several research groups have reported high HER activity for Pt-modified tungsten carbides in acidic solutions [5],[6], but these carbides are phase mixtures of WC, WC_{1-x}, and W_2C. One of the goals of our research is to distinguish between the effectiveness of these various carbide phases as HER catalysts. Initial studies have investigated the activity of phase-pure WC and Pt-WC thin film electrodes for the HER in H_2SO_4. The composition and phase purity of these films has been verified using XPS and GIXRD, and their activity towards the HER investigated using linear sweep voltammetry (LSV) and chronoamperometry. The LSV measurements show a significant improvement in WC HER activity upon modification by one monolayer of Pt, and post-electrochemistry XPS measurements indicate no loss of Pt from the WC surface. Current work is focused on investigating the effect of different coverages of Pt on the HER activity of WC and its comparison to polycrystalline Pt foils.

References

[1] H. Kita, Journal of the Electrochemical Society, 113 (1966) 1095.

[2] D.V. Sokolsky, V.S. Palanker, and E.N. Baybatyrov, Electrochemica Acta, 20 (1975) 71.

[3] M.B. Zellner and J.G. Chen, Catalysis Today, 99 (2005) 299.

[4] E.C. Weigert, A.L. Stottlemyer, and J.G. Chen, Journal of Physical Chemistry C, 111 (2007) 14617.

[5] M. Wu and M. Nie, Journal of Power Sources, 166 (2007) 310.

[6] D.J. Ham, R. Ganesan, and J.S. Lee, Int. J. of Hydrogen Energy, 33 (2008) 6865.

See more of this Session: Nanomaterials for Energy Production and Fuel Cells II
See more of this Group/Topical: Topical 5: Nanomaterials for Energy Applications

Novel Quaternary Platinum-Ru-Ni-i Alloy Based Anode Electrocatalysts for Direct Methanol Fuel Cell

Karan Kadakia, Chemical Engineering, University of Pittsburgh, Pittsburgh, PA

Prashanth Jampani Hanumantha, Chemical Engineering, University of Pittsburgh, Pittsburgh, PA

Nicolaus L. Rock, Mechanical Engineering & Materials Science, University of Pittsburgh, Pittsburgh, PA

Moni Kanchan Datta, Mechanical Engineering & Materials Science, University of Pittsburgh, Pittsburgh, PA

Prashant Kumta, Chemical Engineering, University of Pittsburgh, Pittsburgh, PA

Direct methanol fuel cells (DMFC) are of interest as suitable power sources for consumer electronic devices, remote and auxiliary power units for transportation. The main advantage of DMFC is the use of liquid methanol fuel rather than gaseous hydrogen in proton exchange membrane fuel cells (PEMFC). The ability to store and handle liquid better than gaseous hydrogen makes it a viable energy source for mobile and portable applications. The energy density of methanol and the amount of energy contained in a given volume of methanol is an order of magnitude greater than even compressed hydrogen. However, the currently achieved power densities of DMFC are still significantly below PEM and other fuel cell types. This is mainly related to the relatively sluggish kinetics of the electrochemical reactions occurring on the catalyst layers at the electrode surface combined with the methanol crossover problem to the cathode compartment. There is therefore a critical need to research and develop new and improved anode and cathode electro-catalysts with higher catalytic activity in order to enhance the efficiency and power density of DMFC, while lowering the precious metal loadings to realize the much desired cost savings to match the performance.

The nominal composition Pt-50at.% Ru is considered the most suitable anode catalyst composition yielding the best catalytic activity for use in DMFCs. Improved anode catalysts can be obtained by synthesizing high surface area Pt-Ru based alloy catalysts combined with engineered proper alloy design by modifying the chemical composition of the alloy, and doping or alloying with other elements onto the catalyst structure. Ternary and quaternary alloys with transition metals have been explored as possible alternatives to pure Pt-Ru electro-catalyst. The addition of other non-noble transition metals could help potentially reduced the cost while providing similar electrocatalytic activity. It has also been proposed that addition of other alloying elements could reduce materials loss and particle ripening during cycling. Ni-Zr and Ni-Ti have previously been shown to be electrochemically active and stable under anode conditions. Recently, sputter deposited Pt-Ru-Ni-Zr thin films have been shown to exhibit promising catalytic activity. In the present work, we report bulk synthesis of nanocrystalline Pt-Ru-Ni-Ti based quaternary alloys, for use as DMFC anodes, by a facile wet chemical synthesis route. X-ray results show complete alloying and formation of phase pure Pt(Ru,Ni,Ti) solid solution, and also nanocrystalline nature of the alloy. Alloy materials have been shown to have moderate BET surface areas (~70m2/g). Electrochemical characterization has been performed on a three electrode cell using a solution containing 1 M methanol and 0.5 M sulfuric acid as an electrolyte while also serving as a source of the fuel. Cyclic voltammetry and polarization results indicate that the nano-crystalline Pt-Ru-Ni-Ti alloy of suitable compositions possessing good chemical homogeneity, reveal promising catalytic activity, demonstrating the potential of the Pt-Ru-Ni-Ti based electro-catalysts for direct methanol fuel cells. Results of these studies will be presented and discussed.

See more of this Session: Nanomaterials for Energy Production and Fuel Cells II
See more of this Group/Topical: Topical 5: Nanomaterials for Energy Applications

Improving Oxygen Reduction Efficiency in Direct Methanol Fuel Cells through Structured Catalyst Design

Jinhua Yang, Department of Chemical and Biomolecular Engineering, National University of Singapore, Singapore, Singapore

Jing Xu, Department of Chemical and Biomolecular Engineering, National University of Singapore, Singapore, Singapore

Mark Saeys, Department of Chemical and Biomolecular Engineering, National University of Singapore, Singapore, Singapore

Jim Yang Lee, Department of Chemical and Biomolecular Engineering, National University of Singapore, Singapore, Singapore

We have adopted a rational approach to the design of fuel-tolerant oxygen reduction catalysts for direct methanol fuel cells (DMFCs). Our methodology is to use structured catalysts with different core and shell components. These catalysts may be prepared by simple galvanic replacement reactions which can be scaled up for volume production. Density functional theory (DFT) calculations of changes in the d-band centers of the surface catalytic metals; and the strengths of adsorbed species are used to gain some insights into the relationship between surface structure and catalytic activity of these heterogeneously structured catalysts. We found that carbon-supported core-shell PdM@PdPt catalysts (where M=Ni, Co, Fe and Cr) can provide us with the greatest latitude in activity and selectivity tuning. A volcano-curve was obtained where maximum ORR activities occur around the compositions PdCo@PdPt/C and PdFe@PdPt/C. It is believed that the Pt d-band centers in these catalysts are at values representing a good balance between O-O bond dissociation and the removal of adsorbed OH groups on the catalyst surface. Compared with commercial Pt/C catalysts, the PdM@PdPt catalysts utilize less platinum, and yet display favorable catalytic outcomes in terms of ORR activity and selectivity (methanol tolerance).

See more of this Session: Nanomaterials for Energy Production and Fuel Cells II
See more of this Group/Topical: Topical 5: Nanomaterials for Energy Applications

PdFe Nanorods as Highly Active Electrocatalysts for Ethanol Oxidation in Alkaline

Kamonwad Yangyuenthanasan, Materials Science and Engineering, Michigan Technological University, Houghton, MI

Zhiyong Zhang, Chemical Engineering, Michigan Technological University, Houghton, MI

Wenzhen Li, Chemical Engineering, Michigan Technological University, Houghton, MI

Palladium (Pd) and its alloys have been demonstrated promising electrocatalytic activity towards ethanol oxidation reaction (EOR) in alkaline [1]. Recently, we have successfully synthesized PdFe-NR with high oxygen reduction reaction (ORR) in acid electrolyte (both in three-electrode-cell and single proton exchange membrane fuel cell PEMFC) [2], based on a modified synthesis approach that was firstly reported by Sun et al. [3]. In this presentation we report a convenient organic phase reduction synthesis approach to small diameter PdFe nanorods (PdFe-NR), which show very high EOR activity in alkaline.

The synthesis of PdFe-NR was obtained through chemical reduction of Pd(acac)2 and thermo-decomposition of Fe(CO)5 in oleyamine (OAm) at 160oC. The EOR activity of PdFe-NR was tested in a conventional three-electrode-cell by recording their cyclic voltammetry (CV) curves. The electrolyte was 0.1M NaOH + 0.5 M CH3CH2OH. It is found that the EOR activity of the PdFe-NR-carbon nanotubes (CNTs) (PdFe-NR deposited on CNT covered glassy carbon electrode GCE) exhibits higher mass activity than Pt/C and PtRu/C, although its onset potential of EOR remains lower than PtRu/C. TEM images show that the PdFe-NR has a mean diameter of 2.5 nm and length of 23 nm. Other Pd-based nanocatalysts, such as CNT supported PdNi, PdCo, etc, and more characterizations, such as XRD, TGA, XPS, etc are currently under working in our lab to achieve better EOR activity and attain an insight to the activity enhancement.

References:

[1]Shen et al, Electrochemical Communications, 2006, 8, 184-188.

[2]Li et al, Electrochemical Communications, 2009, doi:10.1016/j.elecom.2009.03.046.

[3]Wang et al. Angewandte Chemie-International Edition, 2007, 46, 6333-6335. ->

See more of this Session: Nanomaterials for Energy Production and Fuel Cells II
See more of this Group/Topical: Topical 5: Nanomaterials for Energy Applications

Nitrogen Substituted Carbon Nanotubes as An Electrode Support for Pt in Formic Acid Electrooxidation

Sankaran Murugesan, Chemical & Metallurgical Engineering, University of Nevado at Reno, Reno, NV

Kirby Myers, Chemical & Metallurgical Engineering, University of Nevado at Reno, Reno, NV

Ravi Subramanian, Chemical & Metallurgical Engineering, University of Nevado at Reno, Reno, NV

Electrode supports play an important role in effective anchoring and dispersion of Pt metal in electrooxidation of fuels such as methanol and formic acid. The electrode supports reduces the sintering of Pt particles and show high influence in stable operation of fuel cells. At present, commercially Vulcan XC-72 carbon has been used as an electrode support due its properties such as high specific conductivity and low double layer charge compared to other carbon electrode supports. Recently, carbon in nanomaterials such nanotubes, nanofibres, and nanosheets have attracted great interest. Such electrode supports demonstrate different structural, physical, and electronic properties by the arrangement of carbon network. The surface properties of carbon nanotubes, for example, can be altered by chemical treatment or substitutional doping of heteroatoms. This work focuses on the effect of nitrogen substitution in carbon nanotubes (NCNT) as a support for Pt anchoring in formic acid electrooxidation. Synthesis of nitrogen containing carbon nanotubes and how they can help in anchoring Pt in preferential orientation is presented. The results of NCNTs for formic acid electrooxidation have been compared with the other commercial electro catalyst (HiSpec 3000) and commercial support Vulcan-XC 72. The results of this work play a significant role in the development of fuel cell electrodes.

See more of this Session: Nanomaterials for Energy Production and Fuel Cells II
See more of this Group/Topical: Topical 5: Nanomaterials for Energy Applications

Electrically Conductive and Super-Hydrophilic Bipolar Plate Coatings Prepared From Mixed Aqueous Suspensions of Graphene Sheets and Silica Nanospheres

Feng Wang, Chemical and Materials Engineering, University of Cincinnati, Cincinnati, OH

Indrajit Dutta, Chemical and Materials Engineering, University of Cincinnati, Cincinnati, OH

Ruidong Yang, Chemical and Materials Engineering, University of Cincinnati, Cincinnati, OH

Mubarak Alazemi, Chemical and Materials Engineering, University of Cincinnati, Cincinnati, OH

Anastasios Angelopoulos, Chemical and Materials Engineering, University of Cincinnati, Cincinnati, OH

Interfacial chemical domain size and distribution can have a significant influence on the electronic properties of composite coatings. In this paper, we demonstrate how the theoretical predictions of the Gouy-Chapman model for electrostatic interactions between nanoparticles may be used to control heterogeneous chemical domain size and distribution at interfaces. Specifically, we demonstrate through high resolution SEM, TEM, and AFM imaging how the ionic strength, dielectric constant, and pH of suspending media may be systematically altered to vary the three-dimensional distribution of hydrophobic graphene sheets (in the form of ~5 nm thick graphite platelets) and hydrophilic silica nanospheres (~60 nm diameter) simultaneously and irreversibly deposited onto metallic surfaces from mixed suspension utilizing a polymeric binder. The process is used to alter the water wetting properties of bipolar plate materials employed in proton exchange membrane (PEM) fuel cells while maintaining low electrical contact resistance. Such composite coatings are shown to yield advancing contact angles of not more than 10 degrees while maintaining an electrical contact resistance below 5 mΩ-cm^2 for ~100 nm adsorbed layer thickness. This result represents an order of magnitude reduction in contact resistance relative to that of pure silica coatings of comparable thickness prepared utilizing chemical vapor deposition.

See more of this Session: Nanomaterials for Energy Production and Fuel Cells II
See more of this Group/Topical: Topical 5: Nanomaterials for Energy Applications

AUTHOR INDEX

AUTHOR INDEX